Engaging with animals

ANIMAL PUBLICS

Melissa Boyde & Fiona Probyn-Rapsey, Series Editors

Animal death
Ed. Jay Johnston & Fiona Probyn-Rapsey

Cane toads: a tale of sugar, politics and flawed science
Nigel Turvey

Engaging with animals

Interpretations of a shared existence

Edited by Georgette Leah Burns and Mandy Paterson

SYDNEY UNIVERSITY PRESS

First published by Sydney University Press
© Individual contributors 2014
© Sydney University Press 2014

Reproduction and Communication for other purposes
Except as permitted under the Act, no part of this edition may be reproduced, stored in a
retrieval system, or communicated in any form or by any means without prior written
permission. All requests for reproduction or communication should be made to Sydney
University Press at the address below:

Sydney University Press
Fisher Library F03
University of Sydney NSW 2006
AUSTRALIA
Email: sup.info@sydney.edu.au

National Library of Australia Cataloguing-in-Publication Data

Other Authors/Contributors:	Burns, Georgette Leah, editor.
	Paterson, Mandy, editor.
Title:	Engaging with animals: interpretations of a shared existence
ISBN:	9781743320297 (paperback)
	9781743320303 (ebook: epub)
	9781743323861 (ebook: mobipocket)
Notes:	Includes bibliographical references and index.
Subjects:	Animals and civilization.
	Human–animal relationships.
	Human–animal communication.
Dewey Number:	304.2

Cover image: painting by Nigel Hewitt, detail of *The Shadower V*, 2007, 100x110cm, mixed
media on canvas, nigelhewitt.com.au

Cover design by Miguel Yamin

Contents

Acknowledgements

Georgette Leah Burns and Mandy Paterson

The idea for this book was born at the fourth biennial Australian Animal Studies Group Conference, 'Animals, people – a shared environment', held in Brisbane in July 2011. We had the privilege of hearing a
number of outstanding presentations at the conference which inspired
us to think more about the human–animal bond, particularly how that
bond is represented in the Australian context. We are grateful to all the
speakers at this conference for stimulating our thinking and starting us
on the journey to produce this book.

Firstly, we would like to thank the authors for their contributions,
for preparing their manuscripts on time and willingly considering the
feedback from our reviewers, editors and from us. Finalising copy and
ensuring copyright for the photos and diagrams that appear in this
book took effort and commitment. We thank them for that and for their
continual encouragement and belief that the book would be finished.

We are also very grateful to Sydney University Press for guiding
us through the process of publishing a book, to Ian Bytheway for
proofreading, and to the Environmental Futures Research Institute for
their financial support. We were delighted to find an image perfect for
our cover, and greatly appreciate the permission from Western Australian artist Nigel Hewitt for us to use it.

We would like to thank the reviewers who accepted the important
task of reading and providing useful feedback to the authors: Darryl
Jones, Natalie Edwards, Deidre Wicks, Matthew Chrulew, Lesley Chase,

Denise Russell, Tess Williams, Yvette Watt, Steven White, Jane Johnson and Andrew Tribe.

Finally, each editor would like to thank the other for their dedication and support throughout the editing process. We hope this book gives the readers as much enjoyment as it has already provided us.

Introduction

Georgette Leah Burns and Mandy Paterson

—

As co-habitants on the same planet, humans and other animal species engage with each other constantly, and in a large variety of ways. Humans tread, deliberately or not, on ants and other insects. We eat animal flesh and rely on animal labour for many essential products. The large-scale negative influence of humans on animals is obvious. As so-called progress continues across the globe we tear down habitats, pollute waterways, introduce species into new areas and breed species to suit our demands. At the same time, we also welcome animals into our homes as pets for whom we feel affection. Humans can be the saviour of other animals, with colossal resources used in saving certain species from extinction, and in rescuing animals during natural disasters. The plight of a suffering animal would only fail to move the most hardened of humans. In short, there are 'some we love, some we hate, and some we eat' (Herzog 2010). But this is not a one-way relationship. Nonhuman animals can kill humans and spread disease, or offer us comfort and companionship. In many contexts symbiotic relationships are created, where humans and nonhumans simultaneously rely on each other for subsistence and survival. Exploring and understanding these complex relationships is the basis for this volume on shared existence.

The very broad field of animal studies is growing and becoming increasingly cross-disciplinary in its nature, both in Australia and internationally. The Australian Animal Studies Group, for example, formed in 2005, holds biennial conferences, launched an online journal in 2012, and has a rapidly increasing membership. This book builds on the success of the essay-collection format in the field of animal studies. We deliberately chose a title that reflected those of earlier works such as *Considering animals* (Freeman et al. 2011) and *Knowing animals* (Simmons & Armstrong 2007), but one that also indicates the desire to actively move forward and further advance this field of contemporary discourse. The multidisciplinary and interdisciplinary nature of the book incorporates work from historians, veterinary scientists, anthropologists, artists, linguists, environmental scientists and ethicists. In combining the uniqueness of our editorial collaboration with the breadth of perspectives in the selected essays, this book widens the scope of human–animal studies collections, especially in the Australian context.

Engaging with animals began life after its two editors were brought together to organise a conference in Brisbane in July 2011. As an academic at Griffith University, Leah is based on a campus in a forest setting, where native and introduced species of flora and fauna are considered from a research perspective in an abstract and esoteric way. As a senior scientist at the Royal Society for the Prevention of Cruelty to Animals (RSPCA), Mandy deals with lost and neglected domestic animals, farm animals and injured wildlife from an animal welfare perspective and in a concrete and pragmatic way. We each engage with animals daily, though in very different ways, and our experiences and commonalities lead us to explore how a more overtly natural scientific view of animals could be combined with a more social scientific approach.

We set about selecting authors whose work embodies a wider understanding of human–animal interactions. Thus, each chapter is unique for different reasons: the species that is examined, the way it is examined, and the theoretical constructs employed. This volume is the result of our engagement and desire to share knowledge on the many different ways humans and other animals share existence. It is a collaboration between a veterinarian and an anthropologist, an animal welfare organisation and an academic institution. The assembled chapters bring together, in a unique way, a number of perspectives from different disciplines that all address issues of human engagement with other animals. This is achieved while maintaining a focus on the Australian context of these engagements.

Introduction

The themes and chapters

Engaging with animals is comprised of 13 chapters distributed amongst three themes offering different approaches to exploring the overarching topic of engagement between humans and other animals. The chapters present different interpretations of how humans and other species share their existence on the planet. These interpretations come from a wide range of disciplinary frameworks. An Australian focus is a further thread linking all the chapters that consider a diversity of species from large carnivorous wildlife, such as dingoes, to controversial feral cats, symbolic donkeys and little-understood krill.

Attitudes, ethics and interactions

The first theme, explored over five chapters, focuses on human attitudes toward and interactions with other animals. Ethical standards and stances in the field of animal studies are also critically examined.

Broadly tackling a topic relevant to many of the chapters, in chapter one Georgette Leah Burns examines the debates around the practice and concept of anthropomorphism, discussing how an anthropomorphic lens may be beneficial to enhance human understanding of and consideration for other species. The chapter combines the concept of an interspecies ethic with a set of ecocentric principles for engagement that together offer a way of restoring connections between humans and other animals.

In chapter two, Nicholas Malone and Ally Palmer explore the call for an ethical grounding in primatology and biological anthropology. They focus on natural field sites and more artificial zoos as zones in which researchers and great apes interact, arguing that even purely observational research contains complex ethical issues. These issues demonstrate the need for researchers to consider the impact of their work.

How do we practise 'good science' with animals? Simone Dennis questions the objectivity of instrumental reason as she explores, in chapter three, the relationships scientists have with rats and mice in their laboratories. Her interpretation of the interview data suggests that attachment to, rather than detachment from, animals is crucial to 'good science' in which human–animal kinship is recognised despite the laboratory remaining a place of ambiguity and contradiction.

When football and sex collide and the media gets involved, you can expect no less than a public furore. In chapter four, Sandra Burr unpacks one such scandal that involved an Australian football player simulating sex with a dog. Her analysis of the event and public opinion encourages us to think about the way we engage with animals and to reconsider traditional boundaries separating human and nonhuman animals.

Very few of us have the opportunity to witness the animals of Antarctica first hand: our awareness of these species is based almost exclusively on text and image. Thus, how these special species are depicted and presented can have profound influences on public perceptions. Employing the framework of charisma, in chapter five Sophie Fern, Kate Nash and Elizabeth Leane use the ABC television show *Catalyst* to explore how two groups of animals, penguins and krill, are presented to the audience. From this investigation we learn a new perspective on encountering the animals of the southernmost continent.

History, art and literature

The next four chapters explore a second theme: the historical engagement between humans with other animals. This is examined by investigating representations of animals in art and literature.

The war-time story of Simpson and his donkey is well known to most Australians. In chapter six, Jill Bough guides us through a re-evaluation of the Simpson-and-donkey legend by focusing on the relationships between the soldiers and the animals at Gallipoli. The chapter addresses the agency of the donkey, the power of its image to represent uncomfortable human emotion and give hope to the soldiers at war and the Australian people at home. Bough also looks at the donkey's role in this historical event in influencing the shared experience of human and nonhuman animals.

Continuing a single-species focus, in chapter seven Amanda Stuart interrogates the relationship between the way the dingo was artistically represented and the way it was treated in colonial times. A very controversial species in the Australian landscape – not unlike feral cats – there are opposing views about the impacts dingoes have on native species. An understanding of the historical interactions between humans and dingoes assists with comprehending the complexities in current relationships.

Anne Taylor also considers visual depiction but concentrates on its role in promoting ethical generosity to all animals, regardless of size or form. In chapter eight she demonstrates the anthropocentric tendencies of historical depictions of life forms that are invisible to the naked eye. The discussion takes us on a journey through scientific drawings of unfamiliar species into contemporary art where depiction of cryptic species is given imaginative and hybrid forms.

The negotiation of similarity and difference is a pervasive element affecting how and why humans engage with different species in different ways. In chapter nine, Sally Borrell explores three works of fiction to illustrate a range of approaches to imagining and voicing the worldview of nonhuman animals. This linguistic anthropomorphism lends language to animals that can provide an effective means of questioning animal subjectivity while at the same time suggesting an emotional life of animals different from that of humans.

Animal and human welfare

The remaining four chapters explore the third theme that focuses on various aspects of animal welfare in Australia.

Considerable controversy currently exists around the topic of feral cats, their role in environmental damage and wildlife predation, and the most humane and ethical ways to control them. In chapter ten, Mandy Paterson explores these topics with specific focus on the use of trap-neuter-return (TNR) as a popular control method overseas and its application in Australia. The chapter concludes that on its own TNR is not likely to provide a solution to the problem across the whole continent, though it could be effective in specific areas.

Media stories were aired recently in Australia about the treatment of livestock, both on and offshore. Sally Healy surveyed 840 people to understand more about consumer awareness, concern and action around current animal farming practices in Australia. Chapter 11 tells us about this survey that showed consumers are concerned about welfare conditions for livestock but find welfare labelling on food products confusing. This negatively influences their ability to make informed choices that improve animal welfare.

In chapter 12, Clare McCausland takes a novel path to exploring a utilitarian argument against industrialised exploitation of farm animals. Combining theoretical works on property rights and the relationship between commodification and utility, this chapter demonstrates that supporters of utilitarianism can make an abolitionist argument against animal exploitation.

Can we know the extent to which animals suffer? In chapter 13 Nicky McGrath and Clive Phillips tackle the topic of emotions in animals, examining the importance of accurately interpreting and assessing animals' feelings. This final chapter explores the ways scientists have attempted to measure emotions and concludes by arguing that more accurate measurement of emotions can improve the experiences of animals under human care.

The interpretations of a shared existence embodied in this book share themes of engagement and geographical location. Combined they demonstrate many different perspectives on how we engage with the nonhuman animals with whom we share this planet, and how they engage with us. Topics of anthropomorphism, ethical consideration and animal rights also permeate many of the chapters, reminding us not only of the dominance of humans on the planet but also of our responsibility to its other inhabitants. There is need to share our resources and find ways to positively exist that can benefit all creatures. We hope these interdisciplinary essays provoke thought about ways forward for mutual engagement, understanding and caring in our shared existence with other animals.

Works cited

Freeman C, Leane L & Watt Y (eds) (2011). *Considering animals: contemporary studies in human–animal relations*. Aldershot: Ashgate.
Herzog H (2010). *Some we love, some we hate, some we eat: why it's so hard to think straight about animals*. New York: HarperCollins Publishers.
Simmons L & Armstrong P (eds) (2007). *Knowing animals*. Boston: Brill.

Part I

Attitudes, ethics and interactions

1

Anthropomorphism and animals in the Anthropocene

Georgette Leah Burns

We live in an era of unprecedented and unsustainable human impact on the earth (Crutzen & Steffen 2003) and there is urgent need to find ways of mitigating and adapting to the environmental changes this is causing. I will argue that, as an important initial step toward dealing with the impacts of this era, humans need to reconnect with nature and particularly with other animal species. I will explain why this is important and how we might achieve such a reconnection.

Our planet faces increasing pressures of globalisation, with rampant commercialisation and consumerism long heralded as the way for humans to achieve economic, and with it social, nirvana. This has served to disconnect humanity from nature, with nature viewed as traditional, backward and characteristically anti-progress (Adams 2003). This historical discourse, with its legacy evident in current practice, needs to be challenged and the field of human–animal interactions offers fertile ground in which to start.

This chapter examines the potential role of anthropomorphism in the Anthropocene. The importance of considering environmental ethics, and a necessary shift away from anthropocentric approaches to living on the planet, are discussed. The concept of interspecies ethics is raised and blended with six principles for engaging ecocentric ethics as a suggested way to guide interactions between humans and other animals.

The Anthropocene

Officially we live in the Holocene period, however, in 2002 Paul Crutzen, a Dutch chemist, published an article in *Nature* declaring the current geological epoch so human dominated that the term 'Anthropocene' should be appropriate (Crutzen 2002). The Holocene is the age of the 'whole' plus the 'new', whereas the Anthropocene is very much intended to denote the age of the humans. Following research in the 1990s documenting human domination of the earths' ecosystems (eg Vitousek et al. 1997), the Anthropocene defines a new era in which humankind has surpassed 'natural' ecological drivers (such as volcanoes, earthquakes and tsunamis) to emerge as a globally significant force capable of reshaping the face of the planet (Clark et al. 2005, 1).

Although the term remains informal (Zalasiewicz et al. 2010, 2228) and the extent of human domination questioned (eg Skrydstrup 2013), the term Anthropocene has gained acceptance in public discourse and academic literature over the past decade, with most debate centring on when the era may have started (Crutzen & Steffen 2003; Steffen et al. 2011) and little argument over the fact that humankind is, and is likely to remain, a major environmental force (Crutzen 2002, 23). As Crutzen noted, the 'daunting task ahead for scientists . . . [is] to guide society towards environmentally sustainable management' (2002, 23).

More than a decade on from the birth of the then provocative term, the notion of the Anthropocene and its impacts evoked in many disciplines and associated discourses are evident. Focusing on the relationships between human and nonhuman species, we find that 'The challenges arising with our new geological epoch – the Anthropocene – call on us to think beyond the human and to consider our species in relation to others' (Roelvink 2012, 1). One way we can do this is through consideration of anthropomorphism.

Anthropomorphism

Anthropomorphism is the use of characteristics defined as exclusively human to describe or explain nonhuman animals (Horowitz & Bekoff 2007, 23). Anthropomorphism 'has long been considered a bad word in science' (Clutton-Brock 2005, 958) but the practice remains enduringly popular (Horowitz & Bekoff 2007, 31) in other discourses. It occurs in children's literature and films, for example, where the human characteristics displayed by other animals are often a crucial element in their popularity. The history of anthropomorphism and current debates about it are worthy of attention because, as I will demonstrate, anthropodenial[1] can be used by humans to conceptually distance themselves from other animal species.

Mithen (1996) suggests that the beginnings of anthropomorphic thinking can be dated to 40 000 years ago when it may have assisted early hunters to predict the behaviours of their prey, an argument supported by Fisher (1996) who similarly suggests an evolutionary explanation for our tendency to anthropomorphise. Despite evidence of historical practice, the label itself only came into existence around the middle of the 20th century with the onset of widespread studies of animal behaviourism, and it quickly became considered a pejorative term (Clutton-Brock 2005, 958). Horowitz and Bekoff (2007, 23) describe the near official consensus in studies of animals' behaviour that anthropomorphising is erroneous and must be avoided.

Despite this very negative association, anthropomorphism has had some very well-respected supporters. Charles Darwin used 'mentalistic terms' (Wynne 2004, 606) and anthropomorphic language (Crist 1996; Taylor 2012, 5) to describe animals he observed. Similarly, when Konrad Lorenz began his studies of animal behaviour in the 1930s, marking the beginnings of ethology as a science, he used anthropomorphic language to relate the ways of animals to the ways of people (Clutton-Brock 2005, 958). Frans de Waal (2001) remains a strong supporter of the view, developed with cognitive ethology, that anthropomorphism does not necessarily disrupt scientific observation and can in fact support the notion of similarity between humans and other animals.

Nevertheless, the arguments for and against the humanising of animals in science persist.[2] Wynne (2004, 606), for example, asserts that

---

1   Anthropodenial, the opposite of anthropomorphism, refers to the deliberate rejection of shared characteristics between people and animals (de Waal 2001).
2   There are also many who write about anthropomorphism but attempt to retain a neutral stance. Epley, Waytz & Cacioppo (2007), for example, develop a three-factor theory about why people anthropomorphise without taking sides in the debate about the merits of it for either humans or nonhumans.

'anthropomorphism is not a well-developed scientific system . . . its hypotheses are generally nothing more than informal folk psychology.' Arguing against a perceived recent move towards greater acceptance of anthropomorphism by some scientists, Wynne maintains 'the reintroduction of anthropomorphism risks bringing back the dirty bathwater as we rescue the baby' (2004, 606). The scientific community's argument against anthropomorphism is usually aimed at its subjectivity in a field where objectivity is of paramount concern,[3] and centres on the presumptive nature of anthropomorphic claims. This sentiment is shared by neo-behaviourist JB Kennedy (1992) who writes about the need to resist returning to the damaging delusions of anthropomorphism. Anthropomorphism is also often attacked for its supposed conceptual and logical transgressions, in favour of the legacy of the pre-Darwinian era in which the biological relatedness of humans to other animals was not recognised (Mitchell 2008, 96). For animal behaviourists, the argument for and against the applicability of thinking about animals in human terms is frequently based on whether or not animals are attributed with having consciousness (see, for example, Wynne 2004).

The traditional scientific mantra that studies must be conducted objectively has a large influence on our engagement with nonhumans, with anthropodenial employed to guard against the possibility of human emotions (through recognising a connection with animals) interfering with neutrality. I argue that discourse on, and decisions about, anthropomorphism should not shy away from emotion and should consider the potential merits of this approach for both people and animals and for their interactions.[4]

Scientific arguments supporting anthropomorphism are usually based on observational evidence, disputed by non-supporters on the grounds that these methods lack necessary objective evidence (Wynne 2007), scientific rigour, and that 'science made too easy is bound to be wrong' (Mitchell 2008, 89). Nevertheless, many convincing argu-

---

3 Recent years have seen a loosening of the strict hold of objectivism yet much contemporary science remains haunted by its legacy (Ginev 2012).
4 The importance of human engagement with, rather than denial of, emotional responses to and connections with nonhumans has also been argued by Milton in her text *Loving nature: towards an ecology of emotion* (2002).

ments are put forward. Horowitz and Bekoff, for example, examining how and why we anthropomorphise, suggest a means to analyse behaviour and see anthropomorphic accounts as 'intelligible and practical' guides to understanding animals (2007, 32). If anthropomorphism can support continuity between humans and nonhumans while providing a practical guide to understanding nonhuman species, then it is useful to explore how we might productively engage with it.

Engaging with anthropomorphism

Daston and Mitman (2005) begin with the premise that although anthropomorphism might be a scientific sin, it can be remarkably useful for human understanding. My claim here is that it might also be useful for other animals and, importantly, for our shared and collective future. I dispute the argument that discovering the truth of anthropomorphic validity is more important than accepting its practice. Mitchell (2008, 89) argues 'specific anthropomorphic theses . . . should be subject to the same rigorous and logical reasoning as any other scientific model', yet perhaps the issue here is with proposing anthropomorphism as a scientific model. We can use it to connect with other animals outside the scientific frameworks without losing any of the positive outcomes that may benefit from this connection.

One should exercise caution, however, and recognise the potential limitations of engaging with anthropomorphism in this way. While Bekoff (2007) applauds the use of anthropomorphism as a tool to make the feelings and thoughts of nonhumans accessible to humans, we run the risk of interpreting these only from a human perspective; that is, the interpretation comes from a reality based exclusively on human values and experience (Bradshaw & Casey 2007). This conceptual limitation may mean we are missing, or at the very least misinterpreting, nonhuman thoughts and feelings. A further criticism of anthropomorphism then is that it has the potential to restrict, rather than enhance, our understanding of nonhumans if we focus only on what they share with us (Tyler 2009; Oerlemans 2007), rendering what may be unique to them unknowable. In this context, Milton (2005) argues very convincingly against the concept of anthropomorphism, suggesting that, because the focus is on humans, 'egomorphism' is a more appropriate term. The challenge is to find a way to engage with anthropomorphism that does not privilege the human.

Instead of asking 'what is wrong with anthropomorphism?' (Milton 2005, 257), perhaps it is more fruitful to focus on what is right about it. Contributors to Daston and Mitman's (2005) book point to the positives of anthropomorphism, arguing that it can promote good health and spirits, enlist support in political causes, sell products across boundaries of culture and nationality, crystallise and strengthen social values, and hold up a philosophical mirror to the human predicament. The latter potential is the focus of this chapter. Interpreting nonhuman behaviour in the same terms as we use for our own behaviour, feelings or emotions enables us to have some empathy with other species and creates a social bond. As Shepard (cited in de Waal 2001, 7) explains, 'Anthropomorphism binds our continuity with the rest of the natural world. It generates our desire to identify with them.' If such identification translates to a desire to learn about, and act ethically toward them, then anthropomorphism can have benefits for both humans and non-humans.

This desire for closeness with the nonhuman world seems to epitomise a typically anthropomorphic view (Curtin 2005, 4), which anthropodenial curtails. Anthropomorphic discourse can enhance the emotional connection with nature for humans and may ultimately assist understanding and respect for nonhumans. Discourses that employ anthropomorphism as a way of constructing views of nature, I argue, can be used as a way forward for the shared future of humans and non-humans.

Franklin (1999) also views anthropomorphism as potentially beneficial for relationships between humans and nonhumans. He suggests that humans are increasingly aware of the extent to which they share their life and world with members of other species and are actively seeking possibilities for coexistence. If he is correct, then the time is ripe for my proposal of discontinuing anthropodenial and striving for coexistence through embracing anthropomorphism. Franklin (1999) links this increased awareness, which has triggered a change in views, to postmodernism. Postmodern understandings of relations between humans and animals are characterised by stronger emotional and moral content. This can be evidenced by the changing face of zoos, which have largely moved away from arrangements where zoos were akin to prisons aimed at separating people and nature and where visitors went to merely gaze and spectacle at an 'other' (Franklin 1999). Increasing awareness of, and sensitivity to, our shared existence with other animal species may also underlie the reasons behind the growing number of animal rights and animal liberation groups, and the increase in recent protests against factory farming and animal experimentation (Best 2006).

Shifting values and ethics

Manfredo (2008) says the increased awareness of a shared existence with other animals leads to a 'mutualism orientation'. Based on results from a study in north America on human value of wildlife, Manfredo et al. suggest that

> [a] wildlife value orientation shift, from dominance to mutualism, is occurring as a result of modernization . . . A mutualism wildlife value orientation . . . views wildlife as capable of living in trust with humans, as life-forms having rights like those of humans, as part of an extended family, and as deserving of caring and compassion. (2009, 39)

People with a strong mutualism orientation are 'more likely to view wildlife in human terms, with personalities and characteristics like those of humans' (Manfredo et al. 2009, 39); that is, they are more likely to engage in anthropomorphic practice. This supports the earlier views of Franklin and, if such a value orientation shift is occurring, suggests a greater willingness by humans to explore alternatives to the ways we engage with nonhumans.

Dawkins (2012) says we engage with animals because they matter to us: they matter because they are of use to us and anthropomorphism should be avoided because it blurs this fact. This is opposite to my assertion that nonhumans matter for reasons beyond their immediate use to humans, and that anthropomorphism can assist with this understanding. Nevertheless, our goals are the same. Dawkins (2012) and I both believe there is a need to rethink our attitudes to nonhumans because they inhabit the planet with us and because a rethink could be beneficial to both human and nonhuman animals. Perhaps Dawkins is more practical and humans will only conserve something if it is of extrinsic value, but I'd like to think not, and I am not alone with this thought. Plumwood (2002), for example, is strongly condemning of instrumentalism, of valuing other species in terms of their use to humans. Her alternative is to call for a decentralising of human-centred ethics, in the form of a dialogical interspecies ethics.

Interspecies ethics

Twenty-five years ago the proposal of interspecies ethics offered a new frontier into human engagement with nonhumans (Rolston 1988). The uptake of it, in academic discourse and in practice, has been disappointingly slow. Although Plumwood explored interspecies ethics in great detail from the perspective of an environmental feminist and it was, encouragingly, introduced as a possible application for animal welfare in veterinary science (Arkow 1998), it seems to have gained little ground.[5] This situation may be a consequence of the legacy of human philosophical engagement with nonhumans, which deserves attention before we can further explore the potential of the interspecies ethic.

Much literary discussion has been devoted to the separation of humans and nature, and humans and nonhumans, by the Cartesian culture/nature dualism. This separation has been exacerbated by the scientific communities' fear of anthropomorphism, and given legitimacy by unchecked notions of anthropocentrism.

Anthropocentrism involves separating ourselves from nature in order to exploit it. This weakens our sense of ecological reality and we lose the ability to situate ourselves as part of nature (Plumwood 2002, 98). This human centredness is fostered by rationalist culture 'to the extent that rationality is taken to be the exclusive, identifying feature of the human' (Plumwood 2002, 98).

---

5   More recently, Haraway (2008) called for humans to start attending to the animals with whom our lives intersect and the notion of attentiveness is often associated with an 'interspecies etiquette' (Warkentin 2010). However, attention and etiquette appear to hold less conviction than a new ethic.

From an anthropocentric viewpoint then, animals are constructed as radically other, and humans are empathically separated from them. Nature is seen as being lower order and as lacking continuity with humans. The focus is on the differentiating, rather than shared, features between humans and nonhumans (Plumwood 2002, 107). As discussed, anthropomorphism's value in this is its opposite focus on shared or perceived shared features. As Plumwood argues, a focus on differentiating features allows for an ethical discontinuity between humans and nonhumans (2002, 107).

There is growing recognition that we need 'to live ethically with others in the Anthropocene' (Roelvink 2012, 3) as authors call for acceptance of nonhumans as 'complex living beings' with whom we share the planet 'rather than as two-dimensional symbols, convenient metaphors, and passive objects of study' (Warkentin 2010, 102). However, modernist binaries continue to 'separate and privilege human life over animal life . . . [which proves] problematic for conceptualizing . . . species interdependence and the possibility of ethical relations between humans and other species' (Roelvink 2012, 4). The positing of binary oppositions of subject versus object, us versus them, human versus nonhuman also facilitates a power structure that elevates the position of humans over other animals (Oliver 2009, 5).

A further role of positing human as the opposite to nonhuman lies in its facilitation of knowing what, or who, humans are, yet this appreciation of the 'the role of others in the constitution of humanity' (Roelvink 2012, 15) is rarely made. Historically, disciplines such as anthropology focused on what made humans different from other species. Anthropological linguistics, for example, posed language as a key feature separating humans from other animals (Hockett 1963) and perhaps this separation has served to make us conceptually dependent upon nonhumans (Oliver 2009) whilst denying human animality (Warkentin 2010, 103). Defining human cognitive abilities as superior to nonhuman ones gives humans the sense of a rightful place of mastery over other species. The assumption of human superiority has resulted in environmental degradation, species extinction and abusive treatments of nonhumans (Warkentin 2010, 103). These things accelerate the perilous state of our planet, affecting its ability to sustainably support all forms of life. Acknowledging interdependence, as well as a conceptual dependence (we are human because they are nonhuman) of humans on nonhumans directs our attention to the question of 'how to share . . . resources and life together on this collective planet' (Oliver 2009, 22).

> The acknowledgement of this interdependence with others unsettles the sense of human superiority and potential mastery over animals . . . it challenges the distinction between animal and human life as both can be seen as intertwined. (Roelvink 2012, 9)

Such 'unsettling' is necessary. Humans have been comfortable and largely unchallenged in these superior roles for too long. The state of the planet is now challenging our sense of mastery over nature, causing us to question its effects and search for alternative paths to a sustainable future.

The role of environmental ethics

Searching for alternative paths to the one that has lead us to the Anthropocene, and bearing in mind Plumwood's (2002) call for interspecies ethics, I looked towards environmental ethics for ways we might learn to live collectively with nonhumans. It seems fertile ground for such alternatives because ethics frequently underpin our decisions and actions (Burns et al. 2011, 179), though certainly not always consciously. 'Ethics are [sic] the practical application into daily living of those beliefs which we strongly value as individuals and as a culture' (Arkow 1998, 194). In Plumwood's dialogical interspecies ethics, an ethical relationship is facilitated by the generation of knowledge created through the role of nonverbal communication between beings (2002, 189). This involves 'reconstructing human identity in ways that acknowledge our animality, decentre rationality and abandon exclusionary concepts of rationality' (Plumwood 2002, 194).

Burns, Macbeth and Moore (2011) propose a set of ethical principles for managing interactions between people and wildlife. The principles offer an approach to managing human interactions with nonhumans in ways that are less human centred (anthropocentric) and more ecosystem centred (ecocentric). Traditional ethics consider human relationships and obligations only toward other humans (Arkow

1998, 197) and such human centred ethics view nature and nonhuman species as possessing meaning and value only when they are made to serve humans (Plumwood 2002, 109). In this anthropocentric framework, ethical considerations apply only to humans because the rational human stands at the centre of knowing and of moral judgment (Warkentin 2010, 103). The principles aim to address this imbalance and I have adapted them here for discussion in the broad context of human–animal interactions, bearing in mind our need and desire to increase connections with the nonhuman world (Table 1.1).

Table 1.1: Ecocentric principles for engaging interspecies ethics (adapted from Burns et al. 2011, 185).

Principle	Description
1. Intrinsic value	Recognises that nonhumans have value in and of themselves, and that this value is independent of their usefulness to human activities.
2. Moral reasoning	Focuses on developing moral reasoning by increasing understanding of other species.
3. Moral obligation	Advocates creating behaviour change through increasing awareness of moral obligations.
3. Precaution	Calls for precautionary action in the absence of proof.
5. Avatar	Stresses the interconnectedness of humans and nonhumans.
6. Reflection	Encourages us to reflect on how our ethical position is constructed and alternative ethical positions have value.

Much human engagement with nonhumans is based on the anthropocentric ethical stance of instrumentalism: that nonhumans hold only extrinsic value. Instrumentalism downgrades, or even denies, independent agency and value of nonhumans (Plumwood 2002, 109). Like anthropodenial, instrumental outlooks block our knowledge of, and sensitivity to, nonhumans. Viewed as being only useful for what they can do for us (for example, as a source of food, clothing or companionship) the dominant notion of human mastery over all other species is highlighted. If nonhumans have intrinsic value, however, then they are worthy of ethical concern and humans have moral obligations toward them (Arkow 1998, 197).

The first principle of *intrinsic value* is intended to extend beyond humans to include other species. It proposes that nonhumans have value in their own right, independent of what they can do for humans (Burns et al. 2011, 186), and that human interaction with nonhumans should reflect this. A chicken, for example, is recognised as being valuable for more than just its use to humans as a source of meat, eggs and feathers. The principle reflects the environmental and holistic prerogative of a shared, collective, future with nonhumans and the recognition of intrinsic value in nonhumans demands re-examination of human interactions with other species that aligns with an interspecies ethics approach.

One of the arguments outlined for engaging with anthropomorphism is that it can increase human understanding of, and compassion toward, other species by establishing an emotional connection. Western environmental ethics and animal ethics have historically rejected emotion as epistemologically invalid in moral deliberation (Warkentin 2010, 104) and the second principle seeks to redress this. Increasing understanding of other species may help to develop *moral reasoning* about nonhumans in a way that is less anthropocentric. It is difficult to consider the interests of other species if those interests are not known. Anthropomorphic practice may assist humans to recognise, or at least to interpret, in a framework familiar and accessible to them, the interests and needs of nonhumans.

Establishing moral reasoning can lead to human behavioural change through the development of a sense of *moral obligation* (third principle). Enabling human awareness of the consequences of their actions can compel them to change their behaviour through a sense of moral obligation (Bamberg & Moser 2007; de Groot & Steg 2007) and concern.[6] Promoting this awareness of consequences to not only the individual species but also the ecosystem that supports them (Burns et al. 2011, 186) should be included in this interspecies approach.

---

6 For detailed examples of and discussion about moral concern for nonhumans see Acampora (2006).

The principle of *precaution* is an important aspect of environmental ethics and very relevant to a notion of interspecies ethics. Removing the need for human superiority, this fourth principle concerns precautionary action and the burden of proof. It essentially states that if an action is suspected of causing harm to animals or their habitat in the absence of scientific consensus, then that action is considered harmful and the burden of proof that it is *not* harmful falls on those proposing the action (Burns et al. 2011, 187). It is about treating nonhumans with respect and affording them the same precautions we would allow ourselves.

'Avatar' is a term used in Hinduism for a material manifestation of a deity, or the descent of a deity to earth (Matchett 2001, 4). This already popular term was thrust into the public arena following release of the successful science-fiction film with the same name in 2009. The *Avatar* fifth principle stresses the interconnectedness of humans and other species as part of shared ecosystems (Burns et al. 2011, 187). This connection can be understood through anthropomorphism which stresses shared features and similarities between species.

An ecocentric perspective as embedded in the principles is not misanthropic (Purser, Park & Montuori 1995). In moving away from anthropocentrism as a guiding ethical position I am not arguing a case for elevating nonhumans to a moral level above humans. Instead the challenge offered, particularly by the Avatar principle, is to conceptualise people as part of nature existing as a component in the natural setting (Burns 2009) and not culturally and environmentally separate from nature (Hytten & Burns 2007).

The success of these five principles depends on the willingness and ability of humans to acknowledge their own ethical positions and to recognise the consequences of their action or inaction. The sixth principle, *reflection*, encourages humans to reflect on the construction of their ethical position and the potential for change over time through interactions with other humans and nonhumans. Understanding complementary and contrasting ethical issues facilitates self-awareness and understanding which is vital to enable individuals to comprehend, and ultimately embrace, alternative ethical positions (Burns et al. 2011, 187).

The six suggested ethical principles for engagement are designed to extend fairness to species other than humans, to ensure that humans in their treatment of other species do not violate the rights of all to flourish (Stenmark 2002). They seek change that will foster a deeper appreciation of the intrinsic valuation of nonhumans, evoke moral reasoning and a sense of obligation through being precautionary in our actions, help us recognise the interconnections between humans and nonhumans, and encourage us to reflect on our individual and collective ethical stances. Ultimately, these six ecocentric principles offer a framework for engagement with other animal species that is less anthropocentric and more holistic. They can facilitate an acknowledgement of animal humanity and human animality by recognising connections rather than focusing on separation. They also serve to afford intrinsic value and rationality to species other than humans, abandoning the traditionally exclusionary concepts of rationality. In these ways, these ecocentric principles may contribute to what Plumwood (2002) had in mind for an interspecies ethics.

Conclusion

If the task of research is to guide society toward a sustainable future (Crutzen 2002), then an important step on that path is to explore the potential for, and possible positive outcomes of, restoring connections between human and nonhuman species. Rather than emphasising differences, we should be embracing similarities and interdependence. However, it is not enough to just recognise interspecies interdependence and thus the need for coexistence. Examination of the Anthropocene, as a consequence of our anthropocentric approach to managing and interacting with nature and nonhuman species, has helped us to get this far. We now need to decide how to respond to this recognition.

Although attributing human characteristics to nonhumans is prolific in some discourses, it is largely avoided in scientific and academic work. I have proposed anthropomorphism as a philosophical mirror to the human condition, allowing us to understand nonhumans and ourselves better but also to understand that the human condition is not just human, we share it with other animal species. I have also suggested we use anthropomorphism as a way to connect, emotionally, with nonhumans and have argued for anthropomorphism to be accepted as a valid form of engagement and potential generator of knowledge without being considered a rigorous scientific framework.

The current world environmental crisis offers opportunity to focus an ethical lens on our interactions with other species. This is why I have proposed we acknowledge our anthropodenial and embrace anthropomorphism as a way of emotionally connecting with nonhumans. It is also why I have proposed adoption of the six ecocentric principles as a way of ethically and holistically reframing our interactions with other species. They provide us with a way of engaging with nonhumans that does not privilege humans and can be achieved through extending intrinsic value to nonhumans, developing moral reasoning and obligation around a sense of a necessarily shared and collective future, taking precautionary measures, and stressing the interconnectedness of all species and their habitats while reflecting on the construction and consequences of our ethical stances. The principles offer a sympathetic extension of the interspecies ethics useful in the broad context of human engagement with nonhumans as we grapple with how to move forward in this era of unprecedented human impact on the earth.

Works cited

Acampora R (2006). *Corporal compassion: animal ethics and philosophy of body.* Pittsburgh: University of Pittsburgh Press.

Adams WM (2003). Nature and the colonial mind. In W Adams & M Milligan (eds), *Decolonizing nature: strategies for conservation in a post-colonial era* (pp16–50). London: Earthscan.

Arkow P (1998). Application of ethics to animal welfare. *Applied Animal Behaviour Science*, 59: 193–200.

Bamberg S & Moser G (2007). Twenty years after Hines, Hungerford, and Tomera: a new meta-analysis of psycho-social determinants of pro-environmental behaviour. *Journal of Environmental Psychology*, 27: 14–25.

Bekoff M (2007). *The emotion of animals.* Novato: New World Library.

Best S (2006). Rethinking revolution: animal liberation, human liberation, and the future of the left. *The International Journal of Inclusive Democracy*, 2(3): 1–24.

Bradshaw J & Casey R (2007). Anthropomorphism and anthropocentrism as influences in the quality of life of companion animals. *Animal Welfare*, 16(1): 149–54.

Burns GL (2009). Managing wildlife for people or people for wildlife? A case study of dingoes and tourism on Fraser Island, Queensland, Australia. In J Hill & T Gale (eds), *Ecotourism and environmental sustainability: principles and practice* (pp139–55). Farnham, Surry: Ashgate.

Burns GL, Macbeth J & Moore S (2011). Should dingoes die? Principles for engaging ecocentric ethics in wildlife tourism management. *Journal of Ecotourism*, 10(3): 179–96.

Clark WC, Crutzen PJ & Schellnhuber HJ (2005). *Science for global sustainability: toward a new paradigm.* CID working paper no. 120. Cambridge, MA: Science, Environment and Development Group, Center for International Development, Harvard University.

Clutton-Brock J (2005). Admitting sympathy beyond species. *Nature*, 434: 958–59.

Crist E (1996). Darwin's anthropomorphism: an argument for animal–human continuity. *Advances in Human Ecology*, 5: 33–83.

Crutzen PJ (2002). Geology of mankind. *Nature*, 415: 23.

Crutzen PJ & Steffen W (2003). How long have we been in the Anthropocene era? *Climatic Change*, 61: 251–57.

Curtin S (2005). Nature, wild animals and tourism: an experiential view. *Journal of Ecotourism*, 4(1): 1–15.

Daston L & Mittman G (eds) (2005). *Thinking with animals: new perspectives on anthropomorphism.* New York: Columbia University Press.

Dawkins M (2012). *Why animals matter: animal consciousness, animal welfare, and human well-being.* Oxford: Oxford University Press.

de Groot J & Steg L (2007). Value orientations and environmental beliefs in five countries: validity of an instrument to measure egoistic, altruistic and biospheric value orientations. *Journal of Cross-Cultural Psychology*, 38(3): 318–23.

de Waal F (2001). *The ape and the sushi master: cultural reflections of a primatologist.* New York: Basic Books.

Epley N, Waytz A & Cacioppo JT (2007). On seeing human: a three-factor theory of anthropomorphism. *Psychological Review*, 114(4): 864–86.

Fisher JA (1996). The myth of anthropomorphism. In M Bekoff & D Jamieson (eds), *Readings in animal cognition* (pp3–16). Cambridge MA: MIT Press.

Franklin A (1999). *Animals and modern cultures: a sociology of human–animal relations in modernity.* London: Sage Publications.

Ginev D (2012). Perspectives on the hermeneutic philosophy of science. *Hermeneia*, 12:107–23.

Haraway D (2008). *When species meet.* Minneapolis: University of Minnesota Press.

Hockett CF (1963). The problems of universals in language. In JH Greenburg (ed.), *Universals of language* (pp1–22). Cambridge, MIT Press.

Horowitz A & Bekoff M (2007). Naturalizing anthropomorphism: behavioural prompts to our humanising of animals. *Anthrozoös*, 20(1): 23–35.

Hytten K & Burns GL (2007). Deconstructing dingo management on Fraser Island: the significance of social constructionism for effective wildlife management. *Australasian Journal of Environmental Management*, 14: 48–57.

Kennedy JB (1992). *The new anthropomorphism*. Cambridge: Cambridge University Press.

Manfredo MJ (2008). *Who cares about wildlife? Social science concepts for exploring human–wildlife relationships and conservation issues*. New York: Springer.

Manfredo MJ, Teel TL and Zinn H (2009). Understanding Global Values towards Wildlife. In MJ Manfredo, JJ Vaske, PJ Brown, DJ Decker & EA Duke (eds), *Wildlife and society: the science of human dimensions* (pp31–43). Washington DC: Island Press.

Matchett F (2001). *Krsna, Lord or Avatara? The relationship between Krsna and Visnu*. London: Routledge.

Milton K (2005). Anthropomorphism of egomorphism? The perception of non-human persons by human ones. In J Knight (ed.), *Animals in person: cultural perspectives on human–animal interactions* (pp255–71). Oxford: Berg Publishers.

Milton K (2002). *Loving nature: towards an ecology of emotion*. London, New York: Routledge.

Mitchell SD (2008). Anthropomorphism and cross-species modeling. In S Armstrong & R Botzler (eds), *The animal ethics reader*, 2nd edn (pp88–97). London: Routledge.

Mithen S (1996). *The prehistory of the mind; the cognitive origins or art, religion and science*. London: Thames and Hudson Ltd.

Oerlemans O (2007). A defense of anthropomorphism: comparing Coetzee and Gowdy. *Mosaic*, 40(1):181–96.

Oliver K (2009). *Animal lessons: how they teach us to be human*. New York: Columbia University Press.

Plumwood V (2002). *Environmental culture: the ecological crisis of reason*. London: Routledge.

Purser RE, Park C & Montuori A (1995). Limits to anthropocentrism: toward an ecocentric organization paradigm? *The Academy of Management Review*, 20(4): 1053–89.

Roelvink G (2012). Rethinking species-being in the Anthropocene. *Rethinking Marxism: A Journal of Economics, Culture and Society*: 1–18.

Skrydstrup, M (2013). Tricked or troubled natures? How to make sense of 'climategate'. *Environmental Science and Policy*, 28: 92–99.

Steffen W, Grinevald J, Crutzen P & McNeill J (2011). The Anthropocene: conceptual and historical perspectives. *Philosophical Transactions of the Royal Society*, 369: 842–67.

Taylor H (2012). Anecdote and anthropomorphism: writing the Australian Pied Butcherbird. *Australasian Journal of Ecocriticism and Cultural Ecology*, 1: 1–20.

Tyler T (2009). If horses had hands . . . In T Tyler & M Rossini (eds), *Animal Encounters* (pp13–26). Leiden: Brill.

Vitousek PM, Mooney HA, Lubchenco J & Melillo JM (1997). Human domination of earth's ecosystems. *Science*, 277: 494–99.

Warkentin T (2010). Interspecies etiquette: an ethics of paying attention to animals. *Ethics and the Environment*, 15(1): 102–44.

Wynne C (2007). What are animals? Why anthropomorphism is still not a scientific approach to behavior. *Computational Cognitive Behavioural Research*, 2: 125–35.

Wynne C (2004). The perils of anthropomorphism: consciousness should be ascribed to animals only with extreme caution. *Nature*, 428: 606.

Zalasiewicz J, Williams M, Steffen W & Crutzen P (2010). The new world of the Anthropocene. *Environmental Science and Technology*, 44(7): 2228–31.

2

Ethical issues within human–alloprimate interactive zones

Nicholas Malone and Ally Palmer

> Among all wildlife-related fields, primatology is a vulnerable and verdant field for emersion of the many and often contradictory effects of interspecies bonding.
>
> Rose 2011, 246

In 1995, anthropologist Nancy Scheper-Hughes called for a 'militant anthropology' that is politically committed and morally engaged. This explicit elevation of the ethical to a primary position over the empirical brings anthropologists face to face with certain responsibilities to their research subjects, and calls into question the idea of the anthropologist as an 'objective observer of the human condition' (Scheper-Hughes 1995, 410). Although this suggestion is potentially problematic on a number of practical and philosophical grounds, we extend the call for an 'ethically grounded' anthropology to the sub-fields of biological anthropology and primatology. In this chapter we pursue this line of inquiry by examining the ethics of research with alloprimates (other primates who interact with people), both in free-living and captive contexts. Specifically, we focus on two sites that have faced relatively little ethical scrutiny: the naturalistic field site and the urban zoological garden. We call these contextual spaces 'human–alloprimate interactive zones'. We discuss several ethical issues arising from primatological research in these settings, and suggest that engagement with these issues is important not only for generating an ethically grounded primatology, but also for reconsidering humankind's relationships with the natural world.

As liminal 'almost humans', nonhuman primates – in particular, the great apes – sit on the philosophical boundaries in Western discourses between human and animal, person and object, culture and nature (Haraway 1989; Marks 2002; Corbey 2005). Corbey (2005, 1) argues that Western discourses surrounding great apes since the 17th century have involved 'an alternation of humanizing and bestializing moves', such that great apes have in different contexts been both deliberately distanced from humanity and brought into the fold of moral personhood. Reflecting this position as boundary markers between humans and the natural world, the study of nonhuman primates is an obvious starting point for a re-evaluation of our relationships with, and treatment of, nonhuman animals, as Rose (2011) also suggests in the opening quote to this chapter.

Such a re-evaluation is timely and urgent due to humans' destructive impact on the natural world. Butchart et al. (2010) examined over 30 indicators of the state of biodiversity and concluded that most parameters associated with extinction risk (eg population demographics, habitat quantity and quality) continue to decline whilst the underlying pressures (eg consumption, pollution, and over-exploitation of resources) steadily increase. Our closest living relatives have not escaped the present biodiversity crisis. Within human-altered ecosystems the world over, approximately half of the known nonhuman primate species face extinction, their fates inextricably linked to the welfare and the activities of humans on both local and global scales (IUCN 2012). These results call into question the adequacy and efficacy of our current engagement with, and response to, the biodiversity crisis. As such, there is an urgent need for humans to rethink their fundamental relationships with the planet's ecosystems in order to reverse these trends. Specifically, Rose (2007) has proposed that humans need to enact a shift from biological dominance to 'biosynergy', or reciprocally supportive affiliations with other species.

We draw on our own research as a way of not only examining primatological research protocols, but also as a way of interrogating institutional and philosophical boundaries between humans and the natural world. We argue that current protocols governing the ethical oversight of research reinforce philosophical boundaries between sub-ject and participant, person and non-person, human and animal. We set out to contest these dichotomies and discuss possibilities for moving towards research protocols that more appropriately reflect the continu-ity of life and an ethos of biosynergy.

Of primatology and primatologists

Research on nonhuman primates can originate from a variety of acade-mic disciplines. Most commonly, primatologists are trained in the tra-ditions of the natural or social sciences, from such diverse fields as bio-logy, anthropology, psychology, medical health sciences, and zoology. Within these fields of study are various degrees to which methodolo-gies are subject to interrogation and reflexive analysis. For example, an anthropologically trained primatologist may place greater value on critical examination of research methods than the psychologist/prim-atologist, who may instead pay greater attention to the statistical signi-ficance of findings. No matter which discipline they originate from, all primatologists are subject to the policies of institutional review boards. Furthermore, as MacClancy and Fuentes (2011, 11) argue,

> seasoned fieldworkers know all too well that moral judgment in the field is less a simple minded application of ethical verities than a con-stant, evolving negotiation of responsibility with all those involved in the research enterprise.

As such, disciplinary traditions matter in shaping how primatologists approach ethics in the field.

A similar diversity exists among the potential sites and subjects of primatological research. Primatologists extend their research gaze onto primates from unhabituated and free-living populations, to captive-bred subjects in biomedical laboratories and many contexts between. In contrast to the contentious public and professional debate on the value of laboratory research involving primates (Longo & Malone 2006), sci-entific research on primates in the field has been subject to less public scrutiny and ethical regulation. This is especially true of 'non-invas-ive' or 'purely observational' research as compared to experimental manipulations involving capture and sedation. We believe that this relative lack of institutional oversight for 'non-invasive' primatological research stems from the general belief in the inherent beneficence of the fieldworker, perhaps best exemplified by Bruno Latour in Strum and Fedigan's *Primate encounters*:

> You [field primatologists] are terrific and without you the primates would not be part of our lives, of our sociology, biology, comparative ethology, ethics, conservation, neuropsychology. They would be confined to pests to be destroyed by natives, to trophies to be hung on the walls of their villas by hunters, and to constrained lab animals just a bit more expensive than guinea pigs. (2000, 535)

The underlying premise of Latour's position is that inherent benefits are derived from primate field research, whether data are applied directly to inform conservation tactics, or when tales are absorbed into society's collective consciousness.

Equating 'knowing more' about free-living primates with 'inherently good' has become an assumption of fieldwork endeavours (Malone et al. 2010). Observational research with captive (including zoo-housed) animals can be justified along similar lines – after all, they are already there. Zoological gardens around the world are epicentres of animal representation and exhibition to millions of visitors annually. Despite a legacy of deplorable confinement conditions, procurement via wild capture and less-than-informed husbandry protocols, some 21st-century zoos act as conscientious participants in the advancement of a sustainable and harmonious relationship between humans and the rest of the natural world. Published mission statements reveal that zoos commonly represent themselves as institutions devoted to the appreciation of natural biotas and the conservation of biodiversity (Patrick et al. 2007). Appreciation and awareness are achieved, according to these statements, via educational, research and recreational opportunities in the zoo setting. Aims of studying zoo animals include, but are not limited to: informing husbandry protocols, developing more effective breeding programs, and training new investigators. Direct and indirect benefits to the animal subjects can be imagined, but 'research' itself typically carries positive connotations. As the topic at hand is ethics, however, we must be prepared to question these fundamental assumptions and research codes of conduct.

The ethics of human–alloprimate interactive zones

Primates are routinely displaced from their habitats, hunted for meat, captured for trade, housed in zoos, made to perform for our entertainment, and used as subjects in biomedical testing. They are also subjects of research inquiries by primatologists. Primatologists have typically valued observations of primates in remote and wild/naturalistic settings, but some researchers have recently begun to focus on areas where humans and nonhuman primates are ecologically and socially entangled (areas that we refer to here as 'human–alloprimate interactive zones'). This new focus on human–alloprimate entanglements within primatology reflects the recognition that very few primate habitats are free from humanity's ubiquitous influence, and requires primatologists to employ new toolkits and theoretical frameworks in their research (Fuentes & Hockings 2010; Fuentes 2012). In these many and varied settings – from the remote field site to the laboratory – primate researchers face a number of important ethical challenges (Fedigan 2010; MacKinnon & Riley 2010).

As with research using humans, studies with nonhuman primates (and other animals) face oversight by institutional ethics committees. We refer to these institutional bodies as animal ethics committees (AECs), although we note that these committees may go by other names, such as animal care committees, or institutional animal care and use committees. Ethical bodies that oversee research with human participants tend to consider a comprehensive range of ethical issues, from participant safety to informed consent, anonymity, and confidentiality. In contrast, many AECs appeal, somewhat narrowly, to the following principles, commonly known as the '3Rs': replacement (using alternatives to animals whenever possible); reduction (using as few animals as possible); and refinement (minimising distress and suffering) (Russell & Burch 1959). Fedigan (2010, 755) notes that for the primatologist working outside of the laboratory in a naturalistic field site or a zoo these criteria may appear 'puzzling and irrelevant'. Some AECs go so far as to permit researchers to identify their projects as exempt from the entire oversight process, if they consider their research to be 'non-invasive' or 'purely observational'.

Similarly, legislation aimed at granting rights to the great apes has tended to focus exclusively on the laboratory setting. For example, the New Zealand Animal Welfare Act of 1999 states that no person can carry out 'research, testing, or teaching involving the use of a nonhuman hominid' unless it has been determined by the director-general of the Ministry for Primary Industries to be either: a) in the best interests of the individual nonhuman hominid; or b) in the best interests of the species to which the individual nonhuman hominid belongs, provided that the benefits to the species are not outweighed by likely harm to the individual (Parliamentary Counsel Office 1999).

Although this represents one important step toward advancing the rights of animals used for research, we aim to broaden the discussion. To this end, we aim here to consider some of the ethical issues not considered by many AECs that accompany research with nonhuman primates. In particular, we focus on two points along a continuum of human–alloprimate interactive zones: the wild and the zoological garden. We limit our discussion to these settings for two primary reasons. Firstly, our own research is based in the wild and in the zoo, so focusing on these settings enables us to draw on personal experiences, both with data collection and with the associated processes of AECs. Secondly, we feel that research with primates in laboratories already faces substantial (though not necessarily sufficient) ethical scrutiny, and hence have chosen to focus on two research zones that are subject to relatively less oversight and critical examination.

Enter into the 'naturalistic' research world

Ethical considerations and the potential for negative ramifications of primate field research are frequently ignored or given only cursory treatment. Field research with nonhuman primates is important for a number of reasons, including assessing the efficacy of conservation strategies and providing information on the impacts of threats (Lonsdorf 2007). Indeed, Wrangham (2000) argues that the establishment of long-term field research is integral to the ongoing conservation of endangered primates. Field research can create a number of problems, however, for both the nonhuman primate research subjects and local peoples. For example, habituated animals are potentially more susceptible to human hunting (in the absence of researcher presence), the presence of researchers can expose nonhuman primates to human diseases, and the cutting of trails and other research activities can alter delicate ecological systems (Köndgen et al. 2008; Travis et al. 2008). Furthermore, primatologists often participate in the economic and political spheres of local human populations (MacClancy & Fuentes 2011). Yet these impacts are often overlooked by both primatologists and by the ethical bodies overseeing primatological research.

To demonstrate the need to broaden discussions of ethics in field primatology, we consider an example of the wide-ranging and highly complex ethical impacts of research with free-living bonobos (*Pan paniscus*) at the Lomako Forest in the Democratic Republic of the Congo. Through the efforts of two local families, in collaboration with foreign research scientists and international conservation agencies (eg the African Wildlife Foundation), a clearer understanding of bonobo behaviour and ecology was achieved (Badrian & Malenky 1984; White 1996; 1986). The activities of foreign researchers provide the local community with short-term economic opportunities, and the intermittent transport of research personnel and logistical supplies provide locals with a certain connectivity to outside goods and services.

Over time, the site developed a reputation throughout the area for its perceived value to the outside world. During the great wars that engulfed the Congo Basin and beyond (1996–2003), the site became a magnet for militant groups. Throughout the period of extreme political upheaval and violence, researchers were largely absent while the local families remained living within a forest of perceived wealth and habituated great apes. Tragically, as a result of soldiers entering the research area, human lives, and a proportion of the bonobo population, were lost.[1] Despite the recent turbulence, the local families continue to welcome the presence of trusted researchers, and the process to formalise the site's protected area status is ongoing.

This history of research and its impact at Lomako exemplifies the grave responsibility and ethically complex landscape that can develop

---

1   Personal communication with Papa Bosco Ikwa, July 2007.

in conjunction with field research into primates. This example demonstrates that assessments of the ethics of field research must involve considerations, compromise, and cost–benefit analyses on a variety of scales: to individual animals, to the neighbours of habituated groups within a population, and to representations of primatological work within local, national, and international spheres. As such, the situation at Lomako is multifaceted and would require in-depth and sophisticated ethical reasoning in any attempt to deem the research as ethical or unethical. Yet it is important to bear in mind that any efforts to assess the ethics of primatological field research are inevitably subjective, and are shaped by cultural understandings of humans and animals, individuals and collectives, and the relationships among these parties.

Given that primatology from the Euro-American tradition is embedded in wider Western discourses – which tend to emphasise boundaries between culture and nature, human and animal – primatologists entering into the complex ethical landscape of the field inevitably carry with them certain personal and cultural assumptions. For example, we suggest that primatologists often tend to relegate the exposure of risk to individuals as 'means to an end', which are justified by an accrual of benefits occurring at an organisational level above that of the individual (eg understanding of the ecosystem, informed conservation of the taxa). As we suggest in the next section, a similar logic of the 'greater good' of conservation outweighing potential harm to individuals is also frequently invoked in support of zoos.

Enter into the artificial world of zoos

As with research in the field, zoo studies tend to be subject to very little ethical scrutiny. Yet like the field, the zoo poses a number of ethical challenges for human engagement with nonhuman primates. Unlike the field primatologist, the researcher in the zoo is but one of many humans observing and interacting with zoo-housed primates. For this reason, the ethical issues with the zoo apply less specifically to the researcher, and rather encompass the complexities and problems represented by the zoo institution as a whole.

Hosey (2005) suggests that the zoo is distinguished from other nonhuman primate habitats by the regular presence of large numbers of unfamiliar humans, space restrictions, and the fact of primates' lives 'being managed'. Although other habitats display some of these features (eg primate pets face space restrictions, temple-living monkeys are similarly exposed to large numbers of unfamiliar humans), Hosey argues that these three features act in concert to create a distinctive zoological garden environment, which influences primate behavioural repertoires and social interactions in specific ways.

One subject of recent attention has been the impacts of visitors on the behaviour and welfare of zoo-housed primates (see Hosey 2008; Davey 2007 for reviews). Although it is difficult to quantify the impact of visitors on zoo-housed primates' welfare and behaviour, research on the subject has documented both negative and positive interactions and effects. On the one hand, zoo-housed primates sometimes appear willing to engage in non-aggressive interactions with visitors, such as those Cook and Hosey (1995) documented with chimpanzees housed at Chester Zoo, England. Ethological research with orangutans at Auckland Zoo, New Zealand, similarly documented non-aggressive, voluntary interactions between visitors and a particular juvenile orangutan (Palmer 2012). On the other hand, a number of studies have suggested that the overwhelming effects of visitors are negative for many groups of zoo-housed primates. Not only is visitor-directed aggression often evident, but intra-group agonistic behaviour may increase with greater visitor presence (Hosey 2008). Although the researcher is but one of many unfamiliar humans observing nonhuman primates in the zoo, it is worthwhile for researchers to bear in mind that their presence may not necessarily benefit their subjects.

Ethical issues with observing alloprimates in zoos do not stop here. A number of authors have taken issue with the philosophical justification and implications of the act of watching nonhumans in the zoo, and have suggested that the zoo problematically presents nonhumans as subjects of human domination and control. Scholars such as Acampora (2005), Berger (1980) and Malamud (1998) have pointed towards an unequal dynamic between those who look and those who are looked at, which renders animals 'absolutely marginal' within the zoo environment (Berger 1980, 24). These authors propose that this 'exploitative consumption of an objectified animal, debases the watcher as well as the watched' (Malamud 1998, 5) such that zoos both reinforce and reflect an unhealthy relationship between humans and other animals based on power and exploitation. In this same vein, authors have drawn on Foucault (1977) to consider the ways in which the zoo's structure and layout act in a similar way to prisons to reinforce divisions between keeper and kept, watcher and watched (eg Acampora 2005; Beardsworth & Bryman 2001; Malamud 1998; Mullan & Marvin 1987).

The notion that zoos serve to reflect and reinforce distance between humans and alloprimates emerged as a significant theme in ethnographic research conducted by Palmer (2012) focusing on caregivers to orangutans at Auckland Zoo. This research suggested that caregivers' efforts to treat their orangutan charges as equals came into conflict with the distance and inequality that inevitably emerged as a product of the zoo's structure and function. One caregiver specifically noted that the built environment of the zoo – in particular, the presence of bars – prevented her from engaging in social relationships where she treated her orangutan charges as equals. Furthermore, distance between caregivers and charges was reflected in caregivers' vision of zoo animals as 'martyrs' for their species, who face restricted freedom for the greater good of conservation (zoo animals act as a 'backup population' for their wild kin) and education (they serve as 'ambassadors' for their species). Caregivers expressed feeling extremely 'guilty' about using zoo animals as sacrifices for the greater good, since this conflicts with their vision of nonhuman great apes as persons deserving respect and freedom. As Regan (1983) suggests, the logic that individuals serve as sacrifices for their species would not be applied to humans – at least, not to law-abiding humans considered to possess their full cognitive faculties and considered to be free of contagious diseases. In this sense, zoos represent a challenge to the treatment and vision of nonhuman primates as moral persons, a perspective which was brought to public attention with the inception of the Great Ape Project in the early 1990s, which aimed to bring great apes into the realm of moral and legal personhood by granting them legal rights (Cavalieri & Singer 1994).

Bridging the gap

It has been our intention to elucidate some of the ethical challenges faced by primatologists working in the human–alloprimate zones of the field and the zoo as these are research sites that consistently face little oversight by institutional ethics bodies. Yet even laboratory studies involving primates face less ethical oversight compared with studies of humans. As Sommer (2011, 38) has pointed out, while researchers of humans go to great lengths to have research approved and must obtain formal consent from participants, 'primatologists can do a lot that cultural anthropologists cannot or should not do'. Sommer (2011) suggests that this state of affairs generates some important ethical questions for researchers who wish to view their nonhuman research participants (eg great apes) as moral persons.

In a sense, this disconnect between research protocols with humans and animals reflects a longstanding tendency in the West to ignore aspects of interconnectivity between humans and animals, and instead emphasise the separation of humans from the natural world. We briefly consider proposals for rectifying this deep imbalance between the ethical oversight given to studies of human versus nonhuman research subjects. In particular, we consider the suggestion that a method akin to ethnographic participant observation might be applied to non-human research subjects (Mullin 1999; Noske 1997; Moore & Hannon 1993; Griffin 1981). Noske (1997, 169) proposes that since anthropology is already a 'science of the Other', the toolkit of ethnography is uniquely positioned to provide insight into nonhuman lives in a way that acknowledges their agency and subjectivity. Although few researchers explicitly link their methods of studying nonhumans to ethnographic participant observation, Moore and Hannon (1993) argue that this kind of approach can be seen in the work of animal behaviour researchers who apply an empathic approach (such as primatologists Jane Goodall, Dian Fossey and Roger Fouts) in order to treat nonhuman animals as social partners whose minds, preferences and thoughts may be interpreted by human researchers.

The suggestion that nonhumans can be treated as ethnographic research participants has been criticised, however, primarily on the grounds that human researchers cannot truly access animals' minds and therefore run the risk of 'transplant[ing] into animal minds the thoughts and feelings we recognise in ourselves, laden as they are with cultural as well as species-specific bias' (Ingold 1988, 9). As Madden (2010, 182–83) argues, although it is self-evident that animals often act as ' "real" social actors' in human societies, treating nonhuman animals as fully fledged ethnographic participants is problematic because we are not, in his view, 'at the point of a human–animal intersubjectivity that allows us to see the world through their eyes' (see also Sommer 2011). As Tapper (1988, 58–59) notes, language is even more of a barrier between species than it is between human groups, since we do not even have animals' native categories available for translation. Others have proposed that animals differ too markedly to humans in their emotional and mental capacities for understanding to emerge. This perspective is perhaps best summarised by Wittgenstein's (1958) suggestion that 'if lions could speak, we could not understand them' (cited in Arluke & Sanders 1996, 41). This argument may hold less sway, however, when discussing the possibility of communication between humans and closely related species such as nonhuman primates.

On the other hand, some authors have pointed out that just as we cannot claim to have insight into animal minds, we similarly cannot ever truly know what other humans are thinking and feeling (Milton 2005; Tapper 1988; Midgley 1988). For this reason, Milton (2005, 265) suggests that we should view our lack of knowledge of animal minds as fundamentally no different in kind to communication gaps between human beings; the only difference is that when observing nonhuman animals the potential for error is greater because of differences in species-specific emotional responses. Thus, Milton suggests that although we are never truly able to understand the subjective state of others – humans and nonhumans alike – we ought to 'still treat the inner world of nonhuman animals as available and perceivable, just as we treat each other's moods as available and perceivable'.

In spite of concerns about our ability to understand animals, communication with nonhumans is clearly possible in some ways, since humans can come to know the behaviour and responses of other species intimately. For example, primatologists and other animal behaviour researchers often come to learn species-appropriate responses and employ such behaviours in their interactions with research subjects (Arluke & Sanders 1996). Thus, although there are indeed barriers to communication between researchers and nonhuman subjects, certain kinds of communication – based on empathy and a familiarity with nonhumans' behaviours – are possible. Furthermore, some have argued that such empathy-based leaps are necessary for enabling ethical engagements and research with animals. From a practical point of view, Bekoff et al. (1992, 474) have argued that the only possible way to attempt to assess pain and suffering in nonhuman animals is to accept the 'inevitability of an anthropocentric stance' and nonetheless carry out assessments of animal welfare, because such investigations are important from a moral standpoint.

More broadly, Smuts (2006) and Haraway (2008; 2006) have both suggested that empathy with nonhuman species is crucial if humans hope to treat nonhumans as minded subjects with whom we can have true intersubjective engagements – which is crucial if we hope to create a healthier relationship between humanity and the other beings with whom we share our planet. Similarly, Fuentes (2006, 125) argues that although it may be difficult to have confidence in our abilities as humans to understand the minds of nonhuman animals – it is not possible to 'think ourselves into the bat' – this barrier does not prevent us from imaginatively exploring animals' personhood: we can ' "be" a bat anthropomorphically and psychologically' and extend the notion of personhood beyond our own species. In this sense, Fuentes (2006) offers hope that in spite of our inability to apply equivalent methodologies for studying human and nonhuman subjects, this does not prevent us from more carefully considering our ethical treatment of nonhumans, and reconsidering the ways in which we use them in our research.

Conclusion

There are deep divisions between research protocols governing studies with humans and nonhumans, particularly with regard to the nature and extent of ethical oversight. Because of the liminal status of alloprimates – particularly the great apes – we chose to examine this division by focusing on the discipline of primatology, a field which Rose (2011, 246) points out is a particularly 'vulnerable and verdant field for emersion of the many and often contradictory effects of interspecies bonding'. Specifically, we chose to draw on our own research with endangered great apes in two human–alloprimate interactive zones: the naturalistic field site and the zoo setting. We have argued that although 'purely observational' research in the field and the zoo is often regarded as inherently good and only minimally problematic, complex ethical issues accompany research in both these settings.

In the field, researchers must consider the impacts of their work on individual study animals and broader ecosystems. They must also consider that their values may compete with those of local people, and must take into account the social and political contexts in determining the impacts of their research. In the zoo, the researcher is but one of many humans observing alloprimates. The presence of large numbers of unfamiliar humans can potentially have negative impacts on the wellbeing of study subjects, and it is important that such impacts be borne in mind by primatologists. Furthermore, we suggest that the zoo can be seen as problematic in terms of the kinds of relationships it fosters between humans and other animals. For example, caregivers to orangutans at Auckland Zoo expressed the idea that zoo animals serve as 'martyrs' for their species, suggesting that individual sacrifice is justified for the sake of the 'greater good' of conservation. Similar ideas are often raised in discussions about the benefits of field research, alongside the notion that 'knowing more' makes such research inherently good. We suggest, however, that ethical issues in the field and in the zoo are complicated, and should be treated with the same degree of attention and delicacy as for research with humans. Furthermore, we note that ideas about ethics are subjective and shaped by cultural traditions and that the Western tendency to erect boundaries between humans and nature permeates discussions of ethics within primatology.

Finally, we considered some potential avenues for rectifying the imbalance between ethical oversight in research with humans and animals. In particular, we considered the suggestion that a method akin to anthropological participant observation might be used to access the minds of nonhumans. Although some have questioned our ability to understand and communicate with nonhumans, it is certainly possible to reach out empathetically and acknowledge the personhood of animals. We suggest that taking these steps might go some way towards generating relationships between humans and nonhumans based on reciprocity, respect and biosynergy. As we have argued, these values are not presently reflected in the current protocols governing primatological practice. We hope that broadening discussions of the ethics of primatology will go some way towards making such a change.

Works cited

Acampora R (2005). Zoos and eyes: contesting captivity and seeking successor practices. *Society and Animals,* 13(1): 69–88.

Arluke A & Sanders C (1996). *Regarding animals.* Philadelphia: Temple University Press.

Badrian NL & Malenky RK (1984). Feeding ecology of *Pan paniscus* in the Lomako Forest, Zaire. In RL Susman (ed.), *The pygmy chimpanzee: evolutionary biology and behavior* (pp275–99). New York: Plenum Press.

Beardsworth A & Bryman A (2001). The wild animal in late modernity. *Tourist Studies,* 1(1): 83–104.

Bekoff M, Gruen L, Townsend SE & Rollin BE (1992). Animals in science: some areas revisited. *Animal Behaviour,* 44(3): 473–84.

Berger J (1980). *About looking.* New York: Pantheon Books.

Butchart SHM, Walpole M, Collen B, van Strien A, Scharlemann JPW, Almond REA, et al. (2010). Global biodiversity: indicators of recent declines. *Science,* 328: 1164–68.

Cavalieri P & Singer P (1994). *The great ape project: equality beyond humanity.* New York: St Martin's Press.

Cook S & Hosey GR (1995). Interaction sequences between chimpanzees and human visitors at the zoo. *Zoo Biology,* 14(5): 431–40.

Corbey R (2005). *The metaphysics of apes: negotiating the animal–human boundary.* Cambridge: Cambridge University Press.

Davey G (2007). Visitors' effects on the welfare of animals in the zoo: a review. *Journal of Applied Animal Welfare Science,* 10(2): 169–83.

Fedigan LM (2010). Ethical issues faced by field primatologists: asking the relevant questions. *American Journal of Primatology,* 72(9): 754–71.

Foucault M (1977). *Discipline and punish: the birth of the prison.* New York: Pantheon Books.

Fuentes A (2012). Ethnoprimatology and the anthropology of the human–primate interface. *Annual Review of Anthropology,* 41: 101–17.

Fuentes A (2006). The humanity of animals and the animality of humans: a view from biological anthropology of JM Coetzee's *Elizabeth Costello. American Anthropologist,* 108(1): 124–32.

Fuentes A & Hockings KJ (2010). The ethnoprimatological approach in primatology. *American Journal of Primatology,* 72(10): 841–47.

Griffin DR (1981). *The question of animal awareness: evolutionary continuity of mental experience.* New York: Rockefeller University Press.

Haraway D (2008). *When species meet.* Minneapolis: University of Minnesota Press.

Haraway D (2006). Encounters with companion species: entangling dogs, baboons, philosophers, and biologists. *Configurations*, 14(1): 97–114.

Haraway D (1989). *Primate visions: gender, race, and nature in the world of modern science*. New York: Routledge.

Hosey GR (2008). A preliminary model of human–animal relationships in the zoo. *Applied Animal Behaviour Science*, 109(2–4): 105–27.

Hosey GR (2005). How does the zoo environment affect the behaviour of captive primates? *Applied Animal Behaviour Science*, 90(2): 107–29.

Ingold T (1988). Introduction. In T Ingold (ed.), *What is an animal?* (pp1–16). London: Unwin Hyman.

IUCN (2012). The IUCN red list of threatened species. Version 2012.2. [Online] Available: www.iucnredlist.org [Accessed 18 September 2013].

Köndgen S, Kühl H, N'Goran PK, Walsh PD, Schenk S, Ernst N, Biek R, Formenty P, Mätz-Rensing K, Schweiger B, Junglen S, Ellerbrok H, Nitsche A, Briese T, Lipkin WI, Pauli G, Boesch C & Leendertz FH (2008). Pandemic human viruses cause decline in endangered great apes. *Current Biology*, 18: 1–5.

Latour B (2000). A well-articulated primatology: reflections of a fellow-traveller. In SC Strum & LM Fedigan (eds), *Primate encounters: models of science, gender, and society* (pp358–381). Chicago: University of Chicago Press.

Longo SB & Malone NM (2006). Meat, medicine, and materialism: a dialectical analysis of human relationships to nonhuman animals and the environment. *Human Ecology Review*, 13(2): 111–21.

Lonsdorf EV (2007). The role of behavioral research in the conservation of chimpanzees and gorillas. *Journal of Applied Animal Welfare Science*, 10(1): 71–78.

MacClancy J & Fuentes A (2011). Introduction: centralizing fieldwork. In J MacClancy & A Fuentes (eds), *Centralizing fieldwork: critical perspectives from primatology, biological, and social anthropology* (pp1–26). New York: Berghahn Books.

MacKinnon KC & Riley EP (2010). Field primatology of today: current ethical issues. *American Journal of Primatology*, 72(9): 749–53.

Madden R (2010). *Being ethnographic: a guide to the theory and practice of ethnography*. London: Sage.

Malamud R (1998). *Reading zoos: representations of animals and captivity*. New York: New York University Press.

Malone N, Fuentes A & White FJ (2010). Ethics commentary: subjects of knowledge and control in field primatology. *American Journal of Primatology*, 72(9): 779–84.

Marks J (2002). *What it means to be 98% chimpanzee: apes, people, and their genes*. Berkeley: University of California Press.

Midgley M (1988). Beasts, brutes and monsters. In T Ingold (ed.), *What is an animal?* (pp35–46). London: Unwin Hyman.

Milton K (2005). Anthropomorphism or egomorphism? In J Knight (ed.), *Animals in person: cultural perspectives on human–animal intimacy* (pp255–71). Oxford: Berg.

Moore DE & Hannon JT (1993). Animal behavior science as a social science: the success of the empathic approach in research on apes. *Anthrozoös,* 6(3): 173–89.

Mullan B & Marvin G (1987). *Zoo culture.* London: Weidenfeld & Nicolson.

Mullin MH (1999). Mirrors and windows: sociocultural studies of human–animal relationships. *Annual Review of Anthropology,* 28: 201–24.

Noske B (1997). *Beyond boundaries: humans and animals.* Montréal: Black Rose Books.

Palmer A (2012). Keeper/orangutan interactions at Auckland Zoo: communication, friendship, and ethics between species. Master's thesis. University of Auckland, Department of Anthropology.

Parliamentary Counsel Office (1999). Animal Welfare Act 1999 no. 142, reprint as at 7 July 2010. [Online] Available: www.legislation.govt.nz/act/public/1999/0142/latest/DLM51206.html [Accessed 18 September 2013].

Patrick PG, Matthews CE, Ayers DF & Tunnicliffe SD (2007). Conservation and education: prominent themes in zoo mission statements. *The Journal of Environmental Education,* 38(3): 53–60.

Regan T (1983). *The case for animal rights.* Berkeley: University of California Press.

Rose AL (2011). Bonding, biophilia, biosynergy and the future of primates in the wild. *American Journal of Primatology,* 73(3): 245–52.

Rose AL (2007). Biosynergy: the synergy of life. In M Bekoff (ed.), *Encyclopedia of human–animal relationships* (pp123–29). Westport: Greenwood Press.

Russell WMS & Burch RL (1959). *The principles of humane experimental technique.* London: Methuen & Co.

Scheper-Hughes N (1995). The primacy of the ethical: propositions for a militant anthropology. *Current Anthropology,* 36(3): 409–40.

Smuts B 2006. Between species: science and subjectivity. *Configurations,* 14(1): 115–26.

Sommer V (2011). The anthropologist as a primatologist: mental journeys of a fieldworker. In J MacClancy & A Fuentes (eds), *Centralizing fieldwork: critical perspectives from primatology, biological, and social anthropology* (pp32–48). New York: Berghahn Books.

Strum SC & Fedigan LM (2000). *Primate encounters: models of science, gender, and society.* Chicago: University of Chicago Press.

Tapper R (1988). Animality, humanity, morality, society. In T Ingold (ed.), *What is an animal?* (pp47–67). London: Unwin Hyman.

Travis D, Lonsdorf EV, Mlengeya, T & Raphael J (2008). A science-based approach to managing disease risks for ape conservation. *American Journal of Primatology,* 70(8): 766–77.

Waldau P (2001). Inclusivist ethics. In BB Beck, TS Stoinski, M Hutchins, TL Maple, A Rowan, EF Stevens & A Arluke (eds), *Great apes and humans: the ethics of coexistence* (pp295–312). Washington, DC: Smithsonian Institution Press.

White FJ (1986). Behavioral ecology of the pygmy chimpanzee. PhD thesis. State University of New York, Stony Brook (NY).

Wittgenstein L (1958). *Philosophical investigations.* Oxford: Basil Blackwell.

Wrangham RW (2000). A view on the science: physical anthropology at the millennium. *American Journal of Physical Anthropology,* 111: 445–49.

3

Of rats, good science and openings to relatedness

Simone Dennis

Acampora (2006) argues that the scientific practitioner operates from the powerful human side of the great divide, wholly detached from the animal object on the other. Further, the animal can never become animate as animal – it is object. In an invocation of Heideggerian language, Acampora (2006, 98) claims that rodent animals in the laboratory are not only physically restrained but are also ontologically reduced to the status of ready-to-hand tools, objects for investigation and examination. Heidegger's (1962) own ontological framework provides few opportunities for the vitality of organic animate being to emerge for serious consideration. While Acampora indicates that he would disagree with Heidegger about animal capacities in time and space outside of the laboratory context, he also indicates that the laboratory is one of the few places where Heidegger was right; herein, animate being fails to ever emerge.

Under the conditions of modern positivist science that Acampora (2006) describes and laments (and in the context of the Judeo-Christian heritage of human supremacy over nature), Bacon's God-scientists appear: humanists who regard the animal as a biological and genetic mirror for self-reflection, as raw material for self-reproduction in a disease-free, improved form (Bacon 1999 [1626]). Echoing Heidegger's (1962) view that under the banner of modernity science itself is arrogated to the place of Plato's 'good' and the Christian God, Acampora details how powerful scientists inhabit the ontotheological domain that the union of science and technology has produced.

Wholly detached from the animals upon whose bodies they operate, these scientists see the fulfilment of Bacon's humanist vision of nature, where nature is made entirely available to the claims and desires of instrumental reason. These scientists equally seem to fulfil Bacon's call to the mastery of nature – through its ontological transformation. This is particularly evident in the production of transgenic animals, as God-scientists here claim not only omniscience but ultimate creative power. As Bacon himself wrote, 'On a given body to generate and superinduce a new nature, is the work and aim of human power' (1999 [1626], 148). Lamenting this state of affairs, Acampora calls for the radicalisation of such relations.

In this chapter, I question these and another of Acampora's claims that

> the laboratory research setting dictates parameters of behavioural operation that desensitise the practitioner to the bodily spectacles enacted under his [sic] experimental surveillance. Indeed, the 'culture' of modern, positivist scientific practice [pivots on] detachment. (2006, 97–98)

I consider 'detachment' in two main ways. First, I question whether or not detachment from the animal is a stance taken by scientists as they encounter animal bodies as research objects in the laboratory. Second, I consider whether modern scientific practice depends on detachment in the sense that the scientist should become 'desensitised' from the animals with whom she/he works. To do so, I make recourse to anthropological understandings of kinship.

Ethnographic data were collected over 12 months (2009–10) with 31 immunologists, virologists and neuroscientists working with either rats or mice in research laboratories in the Australian Capital Territory (see Dennis 2011). My primary mode of data collection was open-ended interviews running over the course of an hour or more, which allowed scientists to comment on their thoughts and feelings about the rodent animals in any way they chose. Participants ranged in age from 35 to 65 years and the group was made up of 16 men and 15 women. The quotations I present herein are a reflection of the broader themes and patterns I identified as dominant in the data as a whole set. As with any interpretation of ethnographic data, I do not make the claim that my interpretation is indisputably right or indeed the only claim that might be made on the basis of the data. Instead I submit it as one interpretation that might challenge some existing assumptions about what goes on in laboratories between animals and humans.

The result of this examination rejects the view that detachment constitutes 'good science' and suggests that attachment is fundamental to practising 'good science'. As I will show in this chapter, however, 'attachment' does not indicate a change in the hierarchical structure of the laboratory, nor does it reduce the lethal violence directed towards animal bodies. Rather, 'good science' is conducted in such a way as to recognise human–animal kinship, and to simultaneously preserve its hierarchical organisation. The laboratory remains a space of ambiguity and contradiction.

Kinship: the opening of relatedness

During my fieldwork I encountered how the conceptual and physical bounds that have long separated human from animal were hierarchically arrayed in the laboratory. Rodents were harmed and killed in the pursuit of scientific discovery, yet my research suggests that such actions were carried out despite laboratory focus on human–animal attachment and similarity, rather than difference and detachment. Animal subjects were understood in the first instance as close (mammalian) gene- and bio-kin. Similarly, I found that it was not uncommon for scientists to consider themselves and all humans biologically and genetically rodent-kin. I also found that some scientists established an ability to understand what rats wanted or did not want to have done to them in the laboratory and rats developed an understanding of human intentions, and could register their refusal, for instance to being handled by a scientist. I explore this later in the chapter.

The co-presence of utility, genetic relationship and interspecies communication indicates a complex and ambiguous situation. Rodents are in some sense our kin and beings with whom communication is possible, while at the same time they are a key piece of laboratory equipment that will be disposed of once its utility is exhausted. This situation calls for an analytic framework that rejects the classic humanist divisions of self and other, mind and body, society and nature, human and animal, organic and technological. These are, I submit, not necessarily characteristic of laboratory praxis and philosophy in postmodernity. Instead, I offer a framework that recognises that hierarchies between animal and human bodies still exist, but also that scientific endeavour is *already* concerned with projects that trouble notions of division and detachment. There is, for instance, as Shanor and Jagmeet (2009) argue, a general willingness to accept that animal life forms are not utterly distinctive from humans (alongside an equal willingness to continue to subject animal life to projects that benefit humans) but these understandings still maintain the existing hierarchy (Wolf-Meyer 2006).

One way of developing a framework that deals with all of these elements is by recourse to anthropological (re)considerations of kinship. As Schneider (1984; 1968) lamented, kinship studies are, overwhelmingly, the expression of anthropologists' ethnocentric biases and disciplinary preoccupations. Biological reproduction, for instance, is taken to be foundational to kinship, a view not shared by all cultures. Post-Schneiderian kinship is characterised by keen reflexive attention to a researcher's own cultural assumptions and practices and the ways in which they might produce particular senses of kinship systems. The study of non-Western kinship systems in and on their own terms has also required investigation of the varying relationships between culture and nature and changing conceptions of nature. This is also true of Western systems of kinship which, in postmodernity, are less sure of how categories of nature and culture might be applied to relatedness. Indeed, the unstable category of nature and its various and unclear relations with culture are now core topics in the study of kinship in Western and non-Western contexts. For example, new reproductive technologies, which have dramatically broadened the possibilities for intervention in and modification of what were once regarded as strictly biological phenomena, have unsettled and even completely destabilised traditional distinctions between nature and culture in the west (see Ginsburg & Rapp 1995; Strathern 1992a; 1992b). Thus, the social construction of science should involve anthropological inquiry into kinship (see Carsten 2004; 2000).

Along with the feminist movement in anthropology in the 1970s and insights from political economy and historical anthropology, the destabilisation of nature and culture that has emerged from post-Schneiderian work has served to open out the concept of kinship to include multiple practices and ideas of relatedness. This has relevance for how we might now examine how science itself is understood and practised in laboratory contexts, how it regards nonhuman bodies and lives therein, as well as how animals and humans might be considered to be related to one another in laboratory (and other) contexts.

The opening of the species border

Rodent animals appear in the laboratory as necessarily ambiguously located between humans and animals; crossing the human–animal border is fundamental to those scientific enquiries concerned with deriving data from nonhuman animal models for application to the human body. For scientific research guided by this intention, nonhuman animal bodies must be sufficiently similar to human bodies for the outcomes of experimentation to have application to human bodies. Biological and genetic sameness are thus critical to the usefulness of rodent animals in the laboratory. The required sameness of nonhuman animal bodies and human bodies is accomplished in and through the subsumption of the species differences of, in this case, rodent research animals, and humans, in favour of a shared mammalian membership based on the close biological and genetic relatedness of humans, rats and mice. These animals appear in the laboratory as human gene-kin and bio-kin. Mice, for instance, on the basis of their close biological and genetic relatedness to humans, are critical models for experimental investigation of human immune diseases without putting human individuals at risk. As one scientist, Paul,[1] explained:

> Mice are much more like us than people [outside the lab] think they are. As mammals, as people are, of course, they have pretty much the same basic body plan, they get the same diseases, they suffer in sim-

---

1 All names of research participants provided here are pseudonyms.

ilar ways to us. For instance, we use the same analgesics to relieve pain in mice that you yourself would have access to in the hospital. Their similarity [to humans] is why we use them.

In Paul's understanding, humans and animals share gene and bio-col-linearity[2] in the laboratory. From this similarity of biology and genes emerges a hierarchy of bodies, wherein one mammalian embodiment is in the service of another. Thus the analytic animal emerges, its ratness or mouseness muted: the species body becomes an unspecific mammalian unit of investigation (see Birke 2003, 207). As Haraway (1997, 89) notes the mammalian homology between the transgenic-breast-cancer-animal-model Oncomouse and people is based on Oncomouse's essence, which, in common with humans, is 'to be mammal, a bearer by definition of mammary glands, and a site for the operation of a trans-planted, human, tumor producing gene'. This mammalian homology is the basis upon which the mouse must be sacrificed, in order 'that I and my sisters [the human mammals in the equation] might live' (Haraway 2008, 76). Such calculations yield sacrificial animals, and are calculations utilising the same formula as Foucault's (2003) calculus of war: the relationship between my life and the death of the other, which enables and justifies the sovereign, and in this case, the scientific, exercise of killing. Such a calculus is developed in line with the thanatophobia – the fear of death – that is fundamental, in Heiddegarian terms, to human existence (see Heidegger 1962). Or, as Leisel, a virologist, explained:

these mice will be sacrificed so that I can find out how to address a problem that is devastating to human health – and definitely a lot of mice die, but it is worth it – imagine if we did no animal research. We would still be beset by polio, for instance, along with a whole raft of other diseases that impact human mammals. It's actually very noble of the mice to give their lives for us.

---

2 'Collinearity' is a term specific to geometry which refers to the property of a set of points lying along a single line. However, the term can be used more generally – as used here – as a synonym for 'aligned'.

Reinert (2007) points out that ideas of sacrifice, such as those that Leisel refers to, are metaphorical and obscure the routinised and technical act of effecting animal death. For Reinert, these banal and routinised actions indicate that the enduring affinity between animals and persons that is central to classical anthropological definitions of sacrifice is lacking in such contexts as the laboratory, as is the capacity of such banal killing to establish a connection between the realms of the sacred and profane by means of a victim that is destroyed in a public ceremonial process. Thus, sacrifice, as Leisel uses it, only indicates that one thing is given up for another, not that a sacrifice in its classical sense is being made. Indeed, for Reinert, these circumstances of death yield the condition 'anti-sacrifice' and indicate the growing 'abyss of essence' that separates the human from the nonhuman and transforms the victim into 'mere equipment, stripped of agency, personhood and other qualities that it might have shared with a human sacrificant' (2007, n.p.).

The location of rodents as gene- and bio-kin may, however, indicate an especial attachment between humans and animals in the laboratory, upon which classical sacrifice may be conducted. For instance, Lynch (1988) argues that the termination of animals in laboratory contexts amounts to more than mere killing. He argues instead that scientific versions of sacrifice contain the key basic themes of sacrifice as they might be understood classically: (1) preparing a victim in such a way as to create and sustain a specific orientation in an abstract space, (2) destroying the victim in order to establish a mediating link between visible and invisible realms, and (3) constituting the victim as a bearer of human attributes. Lynch (1988) argues that animals must in some way become human in order to count as sacrifices – therefore human detachment from animals may not characterise laboratory relations with animals as Reinert (2007) and Acampora (2006) insist.

In Heideggarian terms, I refer here to the possibility of discovering entities (rodents) stripped of their practical or ordinary significance, and provoking a new mode of intelligibility. In laboratory contexts, rats and mice are thematised not as animals, but as mammals, as are humans – something that becomes apparent when rodent flesh becomes data (for instance, in scientific reports, documented findings and so on). This is a transition that requires mice and rats to bear human attributes. The third condition is met when the analytic animal – as opposed to the fleshy animal – becomes a subject of human identification because its anatomical, genetic, and physiological properties are mammalian and therefore human. Thus, for Lynch (1988), and equally in my estimation, the rodent's mammalian membership demonstrates that it is commensurate with its human killer. While he recognises the ways in which humans and animals are hierarchically arrayed in the laboratory, Lynch's assessment of commensurability is one based firmly in the recognition that rats and mice are our bio-kin and gene-kin, and that the great divide between humans and animals is diminished, sufficient that we are the same.

The opening of interaction: fleshy kinship

Alongside the relationship founded on bio- and gene-kinship discussed above, scientists have a relationship founded on interspecies transitivity. This transitivity occurs in both the general sense of transition – that things might move from one state or register to another – as well as in the specific mathematical sense, in that we see a relationship between three elements such that if the relationship holds between the first and second elements, and between the second and third elements, it necessarily holds between the first and third elements. The first element we might say is the analytic animal; that is, the animal that Birke (2003) describes as the data yielding non-specific object that scientists encounter on the job. Encountering the rat or mouse in this way, Acampora (2006) insists, enables the detachment he claims central to modern scientific praxis. The second element might be the naturalistic species mouse, the mouse from which the analytic mouse is derived. The third element is the fleshy animal – the being, human or animal, which is sensing, and sense-able (Merleau-Ponty 1964). Merleau-Ponty (1964) points to the being who understands another being through the commonality of flesh, described as the simultaneous giveness of animality and humanity. Here, Merleau-Ponty suggests that humans already have a feeling for the animal in their own corporeality; since we are both flesh, we have a sense of how the animal might feel, think, behave. Merleau-Ponty uses the phrase 'strange kinship' to capture the sense in which the world is shared among and generally available to the species, despite their evident differences, in the fleshiness of their being. Godway (1998, 50) agrees:

> there is a kinship between the being of the earth and that of my body. This kinship extends to others, who appear to me as other bodies, to animals whom I understand as variants of my embodiment.

It is here that we encounter a quite different relationship with animals in the scientific laboratory than Acampora's (2006) writing allows for. Merleau-Ponty (1964) argues that this feeling for the animal informs scientific endeavour and it is impossible to set aside.

Merleau-Ponty and Acampora, then, are describing two opposite stances: detachment for Acampora, unavoidable attachment for Merleau-Ponty. After Latour (1987) who used the two-headed Janus to introduce ostensibly distinctive and irreconcilable perspectives – Janus talks on the left side like a realist and on the right side like a relativist – I am interested in how both positions are manifest in the laboratory context, and how both are critical to the production of what my research participants called 'good science'. In other words, the first and third elements[3] – the analytic animal and the fleshy animal – are both required for good science.

Brenda, a neuroscientist, conducted what she called 'good science' by deliberately creating close bonds with her six white rats. I asked Brenda about how her affection for her research rats sat with her use of them as analytic animals. She immediately reconciled their status as equipment and animals for which she had affection when she said:

> The way a researcher interacts with animals could, and sometimes does, result in profound behavioural and physiological changes in the animal subject. Things like stress reduction, weight gain – paying attention to them, playing with them – this could be important in understanding how the impacts of some brain disorders could be addressed.

She told me:

---

3 As indeed are the first and second, and second and third elements; thus the mathematical criteria for transitivity are met.

you take it into account in your results – certainly, rats which were stressed out, say from not being familiar with me, could give a different result. I have to interpret a lot by the way they interact with me, about their levels of stress or comfort, but you can tell when a rat is stressed out or doesn't mind having you around.

Brenda seeks deliberately to create certain kinds of bases for data reliability: to eliminate stress as a factor that might change results, for instance. In his rejection of behaviourism, which insists that the scientist be detached from her subject of study, Merleau-Ponty (1964, 93) argues instead that the practising of science requires the scientist to interpret such things as 'rat stress'. Brenda did this by making recourse to what she and the rats shared in common – a fleshy capacity for communication.

During our visit to her rats, Brenda told me 'the rats back themselves into the corners of the cage' making their tails unavailable when she approached them. The rats, Brenda explained, knew she grasped the base of their tail to catch them, and knew how to respond to give Brenda a particular message about their intentions in relation to her move. Thus, Brenda thought the rats were 'telling' her something that she could understand:

> I knew they were refusing me. They really can communicate quite plainly about what they want, and they know what I want when I go for their tails, as lab protocol requires. The rats have not read the lab protocol; they just say 'no' to me. You might think that just means I have to insist, but it is difficult and potentially damaging to them to just grab them – instead, I have to persuade them, by negotiating with them. I might have to give them a treat, or pet them for a bit. It's not just that I impose myself on them – there is a space for negotiation.

Merleau-Ponty (1968) argues that flesh which is constituted by and constituting of the world we each share, ensures our kin-like relations. Our differences in style (rat style, mouse style, human style) distinguish our different experiences. Flesh makes communication possible; we are all sensing and sensible. This kinship neither erases difference nor similarity, makes us neither identical nor separate; for Merleau-Ponty

(1964), animals and people are at once strangers and kin (see Oliver 2007). Brenda's relations with her rats thus existed in an indistinctive zone which required neither complete ratness nor complete humanness to operate; each being was strange to the other, yet their sameness emerged sufficient to offer possible relationships that spanned the animal–human divide. A kind of general contingency between species was recognised. This generality was sufficient to allow for interspecial communication; it was enough to offer the possibility of relatedness and relationship; enough to question the strict situation of rats and mice as biological research equipment.

Such a kinship is constituted and enacted in the thickness of interaction; as Haraway (2008, 4) puts it, 'species of all kinds are consequent on a subject and object shaping [the] dance of encounters'. This dance includes scientific encounters that produce specific rat research subjects and scientific enquirers as indistinct partners, in which rat subjects and human scientists are diminished as bounded categories of being. Dillard-Wright (2009) argues that such a relationship might only seem unusual if we are prepared to accept the exceptionalist argument that human communications differ in kind and in practice from those of other animals.

Precarious life

Despite her kinship with rats – perhaps a kinship of indistinction, perhaps a kinship of fleshy figuration – Brenda killed her rats and entered them into the sacrificial economy of the laboratory, where they were destined to go resultant of their biological and genetic kinship with us. But she entered them 'with unease', as she put it, 'because I genuinely like these little creatures, and have, as I have told you, had conversations with them'. Her comments indicate that she attended to this other kinship – of figuration or of indistinction – with the rats, even as she permanently detached herself from them as spent research objects. Just as Darwin (1871) suggested, experiencing an animal's affection in a research setting haunts the scientist when she or he is confronted by the typical requirements of laboratory work – to wound, to cause suffering, to kill.

As Darwin (1871, 40) noted of the complexities of engaging with animals who could engage back, 'everyone has heard of the dog suffering under vivisection, who licked the hand of the operator: this man, unless he had a heart of stone, must have felt remorse to the last hour of his life'. Loss lies in either direction; should Brenda not kill, the loss is for scientific knowledge, for human betterment. Should she kill, the loss is one that is not just felt because Brenda happens to like rats. Her grief for the rats she sacrifices comes from her knowledge that she is fleshy kin to all ratkind – and it is 'hard', she said, to sacrifice animals 'once you know how alike they are [to us]'. Brenda's knowledge of what the deaths of her rat-kin will produce demonstrates what Butler (2006) describes as 'precarious life'. As Butler also notes, 'One insight that injury [or death] affords is that there are others out there on whom my life depends' (2006, xii). While Butler speaks of the fundamental dependency humans have on anonymous human others, this insight is one applicable to animal others. Butler's is an observation that those at the frontline of human–animal interactions, such as laboratory workers utilising animal subjects, cannot not know.

An ambiguous conclusion

The recognition of rodent–human kinship – manifesting as biological, genetic and/or fleshy relations – is not as radical as it might first appear. Perhaps it should not be totally unexpected that such an indistinctive, unspecied, 'we' comprised of intercommunicating rats with similar genomic profiles to humans should be found in the laboratory. But, as I have shown, the attachments demonstrably present between rats and humans in laboratories – the recognition that bodies similar to those of humans must be treated as though they had the same physical feelings as humans; the notion that rat behaviour is understood through the body of a human, which might not be that different from a rat when it comes to communication – do not destabilise human–animal hierarchies in the laboratory.

Under the banner of 'good science' both human–animal kinship and hierarchy proceed in often complementary ways – as Paul said, 'our [human] similarity to rats explains why we use them'. Often, they feel contradictory, as Brenda recognised as she grieved for the loss of a fellow fleshy communicator. The recognition of the biological kinship between humans and rodents, along with the recognition of interspecial kinship, demonstrates the multiple relations between rodents and humans played out in the laboratory; this makes it very difficult to locate the laboratory neatly in humanist terrain where it has traditionally lain. In the ostensibly unlikely setting of the laboratory, openings exist for strange, indistinctive kinship to develop.

Works cited

Acampora R (2006). *Corporal compassion: animal ethics and philosophy of the body*. Pittsburgh: University of Pittsburgh Press.

Bacon F (1999). The new Atlantis. In R-M Sargent (ed.), *Francis Bacon: selected philosophical works* (pp241–68). Indianapolis: Hackett.

Birke L (2003). Who – or what – are the rats (and mice) in the laboratory? *Society and Animals*, 11(3): 207–24.

Butler J (2006). *Precarious life: the power of mourning and violence*. London: Verso.

Carsten J (2004). *After kinship*. New York: Cambridge University Press.

Carsten J (2000). *Cultures of relatedness*. New York: Cambridge University Press.

Darwin C (1874). *The descent of man and selection in relation to sex*. 2nd edn. London: Murray.

Darwin C (1871). *The descent of man and selection in relation to sex*. London: Murray.

Deleuze G & Guattari F (1987). *A thousand plateaus: capitalism and schizophrenia*. Translated by B Massumi. Minneapolis: University of Minnesota Press.

Dennis S (2011). *For the love of lab rats: kinship, humanimal relations, and good scientific research*. New York: Cambria Press.

Dillard-Wright D (2009). Thinking across species boundaries: general sociality and embodied meaning. *Society and Animals*, 17: 53–71.

Foucault M (2003 [1976]). 21 January 1976. In M Bertani & A Fontana (eds), *Society must be defended: lectures at the Collège de France, 1975–1976* (pp43–64). Translated by D Macey. London: Penguin.

Ginsburg F & Rapp R (eds) (1995). *Conceiving the new world order: the global politics of reproduction*. Berkeley: University of California Press.

Godway E (1998). 'The being which is behind us': Merleau-Ponty and the question of nature. *International Studies in Philosophy*, (1): 47–56.

Haraway D (2008). *When species meet*. Minneapolis: University of Minnesota Press.

Haraway D (1997). *Modest_Witness@Second_Millennium.FemaleMan_Meets_OncoMouse*. New York: Routledge.

Heidegger M (1962). *Being and time.* Translated by J Macquarie & E Robinson. New York: Harper and Row.

Latour B (1987). *Science in action: how to follow scientists and engineers through society.* Harvard: Harvard University Press.

Lynch M (1988). Sacrifice and the transformation of the animal body into a scientific object: laboratory culture and ritual practice in the neurosciences. *Social Studies of Science,* 18(2): 265–89.

Merleau-Ponty M (1968). *The visible and the invisible.* Evanston, IL: Northwestern University Press.

Merleau-Ponty M (1964). *Sense and non-sense.* Evanston, IL: Northwestern University Press.

Oliver K (2007). Stopping the anthropological machine: Agamben with Heidegger and Merleau-Ponty. *PhaenEx,* 2: 1–23.

Reinert H (2007). The pertinence of sacrifice – some notes on Larry the luckiest lamb. *Borderlands e-journal,* 6(3).

Shanor K & Jagmeet K (2009). *Bats sing, mice giggle: revealing the secret lives of animals.* London: Icon Books.

Schneider D (1984). *A critique of the study of kinship.* Ann Arbor, MI: University of Michigan Press.

Schneider D (1968). *American kinship: a cultural account.* Englewood Cliffs, NJ: Prentice-Hall.

Strathern M (1992a). *After nature: English kinship in the late twentieth century.* Cambridge: Cambridge University Press.

Strathern M (1992b). *Reproducing the future: essays on anthropology, kinship, and the new reproductive technologies.* New York: Routledge.

Wolf-Meyer M (2006). Review essay: Agamben's *The open. Reconstruction: Studies in Contemporary Culture,* 6(1). Retrieved on 21 May 2014 from http://reconstruction.eserver.org/BReviews/revTheOpen.htm.

4

Blurred boundaries: humans, animals and sex

Sandra Burr

As companion animals become more firmly embedded in family life, the boundaries separating animals and humans are increasingly blurred. Family pets are afforded the same love, care and consideration as human members of the household and they enjoy privileges that were once only extended to people. Dogs and cats live indoors, they sleep on the furniture and many share their owners' beds. The term 'fur kids' has become an accepted part of the companion animal lexicon and today, more than ever before, companion animals are pivotal to the social fabric of human society. In an age of growing social isolation and increasingly impermanent personal relationships, pets represent the new constant. They are pseudo-humans, substitute family members, faux children for the childless and friends for the socially isolated; for many owners, their pets are the central focus of their lives. Nowadays close relationships with pets are considered normal (Franklin 1999). But how far are we prepared to go in absorbing animals into our lives? Are there some boundaries, including sexual boundaries, that remain not only taboo but uncrossable?

This chapter explores contemporary Australian attitudes towards sexual contact between human and nonhuman animals. It draws on a scandal involving an Australian footballer caught simulating sex with a dog, as well as a number of topical references to bestiality from popular culture including Peter Goldsworthy's controversial novel *Wish* (1995). A number of legal, moral, and social issues emerge that indicate a shift in focus from physical concerns for the welfare of animals to a more nu-anced approach that acknowledges the rights of animals to have their dignity preserved. This change in attitude might lead to understanding that the boundaries traditionally envisaged as separating human and nonhuman animals are dissolving. However, a more considered reading of recent events reveals that anxieties about what it is to be human continue to underpin attitudes towards nonhuman animals and that these remain firmly entrenched.

The event

In late 2010 in Australia, a photograph appeared on a social media site showing a footballer in a compromising position with a dog. The photograph shows a young man lying on his back propped up on one elbow with his pants pulled down and his genitals exposed. He is forcibly pulling a large dog forward and down by its collar, so that the dog's muzzle is in contact with his penis. The man is looking down, concentrating on what the dog is doing. He is smiling. The dog appears to be pulling away.

The image swiftly elicited a blizzard of responses from a broad spectrum of Australian society, almost universally condemning the perpetrator. Opinions varied as to whether a crime had been committed and what, if anything, constituted a suitable punishment for the footballer.

The context of this photograph is relevant. Football in Australia is big business. Clubs attract big-name sponsors and considerable fan bases, and players receive generous salaries. In return players commit to a schedule of daily training, weekly matches, sponsorship obligations and philanthropic work. All codes of football have a long-established culture of binge drinking leading to antisocial behaviours ranging from destroying property and defecating in public, to sexual and physical violence against women.

National Rugby League teams celebrate the end of the playing season with Mad Monday functions. The players dress up in silly costumes and spend the day drinking excessively. Bad things happen on Mad Monday. On 10 September 2010, the Canberra Raiders NRL club held their Mad Monday event on the back of a season ending loss to the West

Tigers team in round two of the finals. By the time a group of players arrived at the home of team mate Josh Miller (who was absent), they had been drinking for several hours and their judgment was severely impaired. One member of the group suggested they play a prank on Miller by photographing his dog performing a simulated sex act with one of them.

Someone who has never been identified sent the resulting photograph to two media outlets. Despite the scandalous nature of the photograph neither media outlet chose to broadcast it. In early November the photograph entered the public domain when it was posted on the fake Twitter account *WyattRoyMP*. The poster later said,

> The reason I decided to post the picture was to highlight the issue of animal cruelty. Whether it be an average Joe or NRL star, what took place in the picture is wrong . . . After nothing was done regarding my concerns I decided to use Twitter to voice my concern. (Moses 2010)

Such was public interest, the photograph rapidly went viral and for a brief time it was the tenth most popular Twitter subject in the world. It spread through internet and social media sites like wildfire provoking a tidal wave of debate about animal cruelty, player misconduct and the reputation of the game.

The fallout

The image was quickly pixelated on the original site, and soon after it was removed altogether. Efforts to suppress the image proved futile as innumerable copies were made and reposted. A private moment of drunken tomfoolery was now firmly in the public domain. Shortly after, Canberra Raider's player Joel Monaghan admitted that he was the man in the photograph. He initially denied responsibility, claiming to be the victim of skylarking by his mates who he said had taken advantage of him while he was passed out drunk. The photograph however, shows him to be fully conscious. Eventually Monaghan confessed to what he described as a simple prank on an absent team mate.

On 6 November Gary Sykes, CEO of Canberra Milk, one of the Canberra Raiders' major sponsors, threatened to withdraw company support for the club unless Monaghan was sacked, although he later recanted and called for disciplinary action instead. He said,

> It's not a good thing when you're thinking that the public is thinking Canberra Milk, a clean-skin company, is associated with people that do this . . . It's not a good look so it's obviously a concern for us, it's bloody awful . . . We don't want this sort of thing associated with our product. (Dutton 2010, 1–2)

Sykes' distress was evident. He could not put a name to what had gone on, describing it as, 'this' and 'this sort of thing', illustrating Ryle's (1994) suggestion that the idea of humans having sex with animals is so offensive that it 'is something we cannot quite talk about properly'. While those who commented on the Monaghan incident were clearly aware that the sex in the photograph was simulated, most found it repugnant. They were not only affronted by the players' lack of judgment, but also by what was perceived by some to a serious breach of animal welfare and/or rights. The code's governing body made it clear that nothing short of sacking would be an acceptable punishment for Monaghan, and on 9 November, almost two months from when the photograph had been taken, and a handful of days since it was made public, Monaghan tearfully announced that he was quitting the club. He never revealed who was with him at Miller's house nor would he say who took the photograph or who sent it to the press and this unwavering loyalty earned him a great deal of public sympathy. The sentiment persisted, however, that his prank with the dog had crossed a line.

At the time of his resignation Monaghan said,

> the reality is this prank isn't going to be forgotten any time soon . . .
> I will have to handle the jokes and taunts . . . The hurt I have caused
> you will haunt me forever. (Moloney & Dutton 2010, 1)

A popular and talented footballer, Monaghan quickly relocated to England to play for the British Super League; however, the transition was not smooth and he suffered widespread humiliation for a long time. Whenever Monaghan was on the field English fans ridiculed him by making barking noises and shouting derogatory comments referencing canines and kennels. The player was given gift bags of dog excrement, although later the bags contained tins of dog food, perhaps indicating a softening of public sentiment toward him (Mascord 2011).

Public opinion

The Monaghan photograph certainly drew a strong response from football fans and other members of the public, who posted on blog sites, wrote letters to the editor and rang talkback radio. While comments about the Monaghan photograph were mostly condemnatory, they were very much divided about the degree of harm done to the dog.

Monaghan's actions in owning up to the deed shifted public opinion. Many felt that while what he did with the dog was wrong he did not deserve to lose his career over it. As local sports journalist Tim Gavel wrote:

> I have received e-mails in the past week from people who blame the media for the demise of Joel Monaghan. Monaghan deserves credit for fronting the media and ringing sponsors to apologise for his actions and accepting responsibility. He paid a high price; in the eyes of some, it was too high given nobody was injured or hurt and Monaghan was contrite. I have always found Joel to be a decent man with a wicked sense of humour. He is always up for a laugh and a prank, but this time he crossed the line . . . Monaghan has learnt the hard way that an indiscretion that takes place in private can quickly become public with enormous ramifications. (2010, 14)

Although not everyone agreed, there was a view that nobody (including the dog) was hurt, that boys will be boys and the incident was just a bit of harmless, if unsavoury, fun. According to Toffoletti in her discussion on footballers and sex scandals with women, it is the sort of assessment that 'frames the modern player as the confused innocent . . . with the presumption . . . being that male sexuality is intrinsic and uncontrollable' (2007, 432). Another view hinted at by Gavel (2010) was that Monaghan's crime was getting caught. In other words the welfare of the dog was not a central issue, but the naivety of the player was. Rebecca

Wilson (2010), writing for the *Herald Sun* took a less charitable view. She described being sickened by the image, calling it depraved and vile, and warned her readers not to look at it.

Public responses to newspaper articles varied. While almost everyone expressed disgust at the photograph, many felt that the matter was not serious because the sex was only simulated.

> It was a drunken prank, not the actual act, get over it. The image is appalling for what it actually simulates, but the act never happened. (Anna. *The Canberra Times*, 6 November 2010, p2)

> Why are people calling this animal cruelty and a criminal offence? Nothing actually happened . . . It was a setup photo meant to resemble the act. Sure sick and disturbing maybe. (Jane4. *The Canberra Times*, 6 November 2010, p2)

Others were concerned for the dog, arguing that its welfare, including its dignity, was compromised, while others were concerned that their children should not be exposed to what they viewed not only as cruelty but disrespect for animals. Indeed, the notion that nonhuman animals deserved the same treatment as human animals in this regard was a common theme. Indeed, one respondent (see below) felt that not to do so posed a threat to the stability of society.

> It's ridiculous to state that just because the dog didn't suffer and didn't understand what was going on that no harm was done. That this was all a bit of harmless fun . . . All three of my boys are big fans of the raiders players and I don't want them to think this kind of disrespectful behaviour and animal cruelty is ok. I've seen the images, and it is NOT harmless fun. ('Johnno, Canberra' in Barrett & Dutton 2010)

> No animal, human or otherwise, should be treated with anything but respect – no exceptions. As soon as we start excusing this behaviour our society spirals downward, because where can you then draw the line? . . . Oh, and let's hope the RSPCA steps in and saves that dog from its current home, because they aren't responsible pet owners. ('steppingsoftly, Melbourne' in Barrett & Dutton 2010)

Some respondents included suggestions that women were in need of protection from the predations of footballers and the inclusion of women in this debate alongside children and animals can be explained by the number of sex scandals involving footballers in recent times (Toffoletti 2007).

> Monaghan is a decent young man who was affected by alcohol. No one was harmed, nobody was hurt and what he did was not as bad as assaulting a woman – worse things have been done to women. (Stephen. 666 ABC Local Grandstand, 5 November 2010).

This brief survey indicates that bestial acts are condemned in the strongest language; they elicit strong feelings of revulsion and antipathy; they evoke concern for the physical and mental wellbeing of the animal involved which is seen as unwilling participant, and bestiality is construed as uncivilised with the potential to destabilise society. It is also evident that there is some confusion about what constitutes bestiality. Is simulated sex with between human and nonhuman animals bestiality? Is coercion a defining factor? Or does something else have to happen; 'does something have to be inserted somewhere?' (Dekkers 1994, 149) before boundaries are crossed?

Bestiality

As Bollinger & Goetschel (2009) argue, bestiality is a complex issue. It is has various definitions including 'sexual relations between a human being and a lower animal'[1] – which is less than helpful in determining exactly what 'sexual relations' are – to the more specific 'sexual intercourse between a person and an animal'.[2]

Contemporary scholars offer a more precise, nuanced interpretation and are at pains to explain the difference between platonic animal love and sex with animals. In 1994 Martin Dekkers suggested that 'bestiality' legitimately described a love of animals because

---

1 See www.merriam-webster.com/dictionary/bestiality.
2 See http://oxforddictionaries.com/definition/english/bestiality.

if you drop the requirement that for sexual contact something has to be inserted somewhere and that something has to be fiddled with, and it is sufficient simply to cuddle, to derive a warm feeling from each other, to kiss perhaps, at times, in brief to love, then bestiality is not a deviation but a general rule, not even something shameful, but the done thing. After all, who does not wish to be called an animal lover? (149)

However, 15 years later Bolliger & Goetschel (2009) recommend substituting the more scientifically correct 'zoophilia' for 'bestiality' on the grounds that

it refers to a strong, erotic relationship with an animal, in such a manner that it leads to its inclusion in sexually motivated and targeted acts, with the direct intention of sexually arousing one-self, the animal or another party. (24)

There is no requirement for penetration to occur; the emphasis rests on intent to cause sexual arousal. By this reading Monaghan's attempt to simulate sex with a dog was clearly an act of zoophilia, but what is less evident is why the Australian public reacted as they did.

Miletski (2009) points out that bestiality has a long history in human society although it is a highly contested issue. In modern times there are prohibitions against bestiality in many countries reflecting a public antipathy towards such behaviour. This antipathy is evidenced in a recent a survey of American adults that revealed their strong disapproval of sexual contact between human and nonhuman animals (Vollum et al. 2004) a sentiment that is further supported by Beirne (2000, 314) who describes bestiality as 'an unusual social practice that is traditionally viewed with moral, judicial and aesthetic outrage', and Ryle (1994, n.p.) when he observes that 'Even in the current atmosphere of doctrinaire tolerance, where all kinds of sexual practices have become widely accepted, the ridicule and opprobrium that attach to bestiality remain'.

The Monaghan photograph was deeply unsettling. His titillation was obvious and the image raised several disturbing unanswerable questions: what if pushing the dog's nose into his genitals signalled a desire for 'real' sex with the dog? Did something more occur that was not photographed and if so, who participated? Were deeper carnal desires at play? Ryle (1994, n.p.) asks why, in this age of compulsive transgression, when no one publicly argues for the right to have sex with animals, do humans find the idea of sexual relations between humans and animals so alarming?

Fudge (2008) offers some insight into this conundrum when she suggests that bestiality breaches the boundaries that separate human and nonhuman animals. She says, 'One form of security we give ourselves in the world is, after all, in the firmness of the boundaries we erect; inside is not outside; human is not animal; self is not other' (2008,18). Bestiality disrupts our sense of identity because we 'have a view of ourselves as fundamentally different from, and superior to, nonhuman animals' (Levy 2003, 450). Morriss (1997) explains this deep-rooted fear very clearly when he says,

> My suggestion is that there is abhorrence, not because it degrades animals, but because it upgrades them – it treats them as something better than they are . . . For a human to have sexual intercourse with an animal implies that the animal is of equal standing to the human. It denies a hierarchy in which animals are lower than humans. (271)

Could this be what respondent 'steppingsoftly' meant in linking bestiality to a downward spiralling of society? Public perceptions of bestiality are that it involves crossing a somewhat nebulous barrier from human to animal and this is a position from which, as Joel Monaghan has discovered, it is difficult to recover. Levy (2003) argues that there are profound consequences for crossing such a barrier.

> to the extent that someone who engages in bestiality, she will find it harder to retain a grip on her identity as a full member of our community, and we will find it harder to admit her to full membership. It is because bestiality is identity-threatening in this way, I submit, that we suspect those people who decide to cross this boundary of psychological illness. (454)

While no one is suggesting that Joel Monaghan was mentally ill, he was adversely affected by alcohol and his ill-judged interaction with the dog resulted in the death of his career in Australia.

In September 2012 Senator Cory Bernardi experienced a similar fate when, as a consequence of comments he made conflating gay marriage and bestiality, he was forced to resign as shadow parliamentary secretary to Liberal leader Tony Abbott. Bernardi offended a significant cross-section of society when he said, in relation to proposals to legalise gay marriage,

> There are even some creepy people out there, who say that it's OK to have consensual sexual relations between humans and animals. Will that be a future step? In the future will we say, 'These two creatures love each other and maybe they should be able to be joined in a union'. I think that these things are the next step. (Bernardi cancels London speech after bestiality remarks 2012, n.p.)

In describing bestiality as a social aberration, Bernardi confirmed the prevailing public sentiment; however, linking it to homosexuality was a step too far. Bernardi's remarks were immediately described by colleagues as 'ill-disciplined', 'hysterical' and 'offensive' (Packham & Kerr 2012), and shortly after he resigned from his position as shadow parliamentary secretary. This was not the end of the matter. While on a flight to London, where he was scheduled to address the European Young Conservative Freedom Summit at Oxford University, the British Prime Minister publicly distanced himself from the Australian senator. Citing concerns that his attendance would be a distraction, Bernardi decided not to address the Oxford summit (Packham & Kerr 2012, n.p.).

Animal welfare/rights

Many of the comments on the Monaghan incident revealed deep concern for the welfare of the dog. Indeed, prevailing public opinion contends that humans have a responsibility to protect nonhuman animals. While animals are theoretically protected by law in Australia, organisations like the RSPCA play a significant role in advocating on their behalf. In their Five Freedoms for Animals manifesto (RSPCA n.d.), four of the five points address physical and psychological harm through freedom from discomfort; from pain, injury or disease; to express normal behaviours; and from fear and distress.

The use of an animal for human sexual pleasure arguably breaches all of these freedoms: without shared understanding animals cannot give their consent to being used for sex and inevitably suffer discomfort, pain, injury, fear and distress when forcibly used for this purpose. Bolliger and Goetschel (2009, 35) claim that sexual practices with animals are more common than generally assumed, but in most countries these sorts of activities are not unlawful unless there is proof of cruelty. As many people observed at the time, bestiality was not a crime in the Australian Capital Territory (ACT) and Monaghan had broken no laws. Michael Linke, CEO of RSPCA Canberra, called for this omission to be rectified on the grounds that existing laws failed to protect the dignity of animals. He said,

> Whilst it appears no laws have been broken, I am disgusted at the image. Every day RSPCA works to improve the bond we have with our pets. The act degrades and shows complete disrespect for animals. This type of image violates the trust we receive from our pets and it is an act that should be made illegal in the ACT immediately. (Doyle 2010, n.p.)

In response, the ACT government agreed to pass an act making it illegal for humans to engage in any sexual activity with animals. Attorney General Simon Corbell said,

> The ACT now joins South Australia as the only Australian jurisdictions to criminalize all sexual activities [with animals], not just the penetration by or of an animal. (Doyle 2010)

While there are laws that legislate against animal cruelty, they do not protect the dignity of animals. Morriss (1997) argues that it is impossible to legislate against offending the dignity of animals because so much of what we do with and to them is offensive, cruel and undignified. Bollinger and Goetschel (2009) argue however, that animals should have the legal right to be treated with dignity and respect.

> This includes, for example, protection from humiliation, excessive exploitation and interference with an animal's appearance, as well as the restriction of certain kinds of contacts with animals which are

not linked to obvious injury, but which concern other animal interests and are to be respected by mankind. According to this view, one important aspect of the dignity of the animal is their sexual integrity . . . The dignity of animals is thus not only injured by violent sexual acts, but any zoophilic act that does not respect the intentions of the animal, and therefore is effected [sic] by using some form of force. (39)

From this standpoint it is probable that the dignity of the dog involved in the Monaghan incident was violated. While we may reasonably assume from the evidence in the photograph that she was not sexually penetrated, whatever happened was not consensual. The dog was resisting and appeared distressed. The reality is, no matter how we choose to view our relationships with companion animals, the bond we have with them is always an unequal one. Furthermore, it is worth noting that while humans may engage in sexual activities outside their own species, under normal circumstances nonhuman animals rarely do.

To return to Bolliger and Goetschel (2009, 42–43),

It is generally recognized, that the sexual freedom of an individual ends where the right for self-determination of another begins . . . The touching of human genitalia by the muzzle of an animal, for example, is not necessarily an animal welfare problem per se and often is only instinctive behaviour. However, if a person systematically exploits such behaviour patterns, then the actions cannot be reconciled with the dignity of the animal.

Without more evidence it is impossible to know if the dog was systematically exploited by footballers, but what is significant is that so many members of the public were sufficiently disturbed by the image to voice their concerns. Without shared language we will never truly know if the dog was disgusted or humiliated but the image of her pulling away from the footballer is evidence that she did not consent.

It is tempting to anthropomorphise the dog in order to gain some deeper understanding about her state of mind at the time of the Monaghan incident; however, Fudge (2008) claims there are dangers in anthropomorphising human–animal relationships. She suggests that when we attribute human qualities and emotions to an animal, our status as humans and that of animals becomes less certain, adding yet another layer to the previous discussion about how relations with nonhuman animals constantly disrupt our notions of what it is to be human.

> Humanist humanity is undone when the animal mind is contemplated, either because the animal mind is revealed to be just like the human mind, thus destroying notions of human superiority; or because the animal mind is recognized as being always beyond our understanding, thus revealing how limited that understanding actually is. Whichever way you approach the issue, what is revealed are the frailties of the human. (52)

Human–animal relations in literature

Human–animal boundaries, particularly sexual boundaries, are theoretically always in place, and they are often tested in both real and imagined worlds. In popular culture imaginative storytelling allows us to consider taboo subjects such as bestiality in an abstract way. In challenging the imaginary divide, Fudge (2008, 10) contends that fiction writers offer 'a way of bridging the chasm that seems to separate humans from animals'. Indeed, creative practitioners appear to be unafraid to test current thinking about bestiality as the following cartoon, television episode and work of fiction show.

Andy Friedman confronts bestiality head on in his cartoon 'My wife! My best friend' that appeared in the September 2012 issue of *The New Yorker* (Figure 4.1). Friedman drew a boxer dog surprising his 'wife', a poodle, in bed with his 'best friend', a man. The dogs are portrayed as pseudo humans and given human values, feelings and emotions and yet the premise of the cartoon is ridiculous because the situation is absurd. By ridiculing bestiality Friedman renders this sensitive and confronting topic safe to think about.

"My wife! My best friend!"

Figure 4.1 Andy Friedman (2012). 'My wife! My best friend!' [Cartoon] *The New Yorker*, September. Used with permission from The Cartoon Bank: A New Yorker Magazine Company.

This jocular finger-wagging is also evident in an episode of the first series of *Rake* (2010) in which lawyer Cleaver Greene is called on to defend family friend Dr Bruce Chandler when an incriminating DVD becomes public showing Chandler and his wife in a three way sex romp with the family dog. Greene is mildly alarmed and resorts to ribald comments about the wife being willing to 'take it with a Rottweiler' to cover his unease. Chandler attempts to normalise the situation by declaring his love for both his wife and the dog, claiming that their sexual encounters are simply an expression of that love. With no footage of the threesome shown, bestiality is dealt with in a safe, abstract way. While it was a little surprising to discover bestiality on prime-time television, it was in accord with *Rake*'s propensity for engaging with topical and controversial subjects such as the Monaghan exposé that had occurred only a month before the episode went to air. *Rake* is primarily a comedy and the Chandler episode was not out of character with other topics covered in the series. However, framing bestiality in a legal context (albeit a fictional one) works to reassure viewers that this controversial subject is being treated seriously. Furthermore, mounting Chandler's predilection within the discourse of love invites viewers to empathise with a man whose prime concern is not lust, but love. Chandler, like the footballer is presented as a victim of his own frailty.

Public antipathy toward bestiality means that engaging in sex with animals is fraught with danger and perpetrators, as we have seen in the Monaghan affair, face humiliation, loss of employment, expulsion from society and, in some cases, criminal charges. In his 1995 novel *Wish*, Peter Goldsworthy explores the consequences for man and beast when sexual boundaries are transgressed. The protagonist, JJ is a man disaffected with life. Through his ability to communicate using sign language, JJ develops a connection with Wish, a young gorilla who is an ex-laboratory animal with surgically enhanced intelligence. When JJ is engaged by Wish's adoptive 'parents', to teach Wish how to sign, the gorilla develops a romantic attachment to him. Recognising that Wish's life is as hopeless as his own, and realising the impossibility of Wish ever finding a suitable mate in her present circumstances, JJ succumbs and has sex with her. When the pair is discovered, JJ is reviled and spurned by everyone he knows and Wish is sent away to a zoo where she commits suicide.

In fictional worlds interspecies love affairs between two consenting beings, such as a man and a gorilla, are made possible. While Goldsworthy fictionally erases the differences between the human and the nonhuman, he is unable to envision a world where such a liaison would be tolerated and the outcome for JJ and Wish is predictably and inevitably wretched. In killing Wish and ostracising JJ, the author reminds us that interspecies sex goes against the natural order. Goldsworthy makes no attempt to contemplate what might happen if the love affair between man and ape had been allowed to continue – a cross-species pregnancy perhaps, resulting in the birth of a mutant gorillorised humanoid, a creature that might conceivably straddle the human–animal divide, is beyond his imagining.

Peter Singer (2001) says that while 'we are great apes', '[t]hat does not make sex across the species barrier normal or natural . . . but it does imply that it ceases to be an offence to our status and dignity as human beings'. Singer's view is an interesting one, and while it tests Morriss' claim (1997, 269) that in the metaphysical sense gorillas threaten our conceptual schema because they are 'like us but not of us' and Fudge's (2008) arguments about the species barrier, it appears to lend some credence to Goldsworthy's (1995) construction of a love affair between a human and an ape. Helen Tiffin (2009, 43–44) argues that 'imaginative writers who wish to re-configure human/animal relations face a number of challenges . . . one of which is how to represent an animal in a medium which is inescapably human-generated'. Goldsworthy addresses this by investing Wish with human emotions and feelings, and while it is generally believed that animals are capable of grieving even to the point of death, in making Wish complicit in her own demise Goldsworthy raises a profound question: if an animal can cognitively contemplate suicide, what does that say about the human–animal divide?

Conclusion

In Australia today bestiality continues to be socially unacceptable and incidents like the Monaghan affair serve to test and reinforce longheld taboos against sex with animals. Few regard sexual contact between humans and animals as benign and bestiality remains high on the scale of undesirable antisocial human behaviour. At one level the debate is concerned with protecting nonhuman animals from physical harm, although, as we know it is still legally acceptable to abuse and mistreat animals through such practices as factory farming, live exports and laboratory testing. While relationships between human and nonhuman animals are by definition unequal, with humans holding the balance of power, there is evidence of a groundswell of well-intentioned desires to redress existing imbalances and injustices.

If we accept that all living creatures deserve to be treated with respect and dignity – and this is a relatively new thread in the animal welfare debate – then the issue of equality between the species must be addressed. It is a path is fraught with difficulties. One fundamental question confronting us as we work through the complexities is how to conceive of and grant true equality to animals in ways that preserve our humanity and their animalness. Clearly having sex with animals is not a measure of equality. Rather, bestiality serves to reinforce our differences and remains, quite rightly, a boundary that must not be crossed.

Love in all its manifestations is the glue that holds society together, but as nonhuman animals become increasingly embedded in family life our understanding of what constitutes love begs the question: is it possible to love animals without completely disempowering them? Events such as the Monaghan affair present us with opportunities to think about our relationships with animals and what it means to share our lives with them.

Works cited

Barrett C & Dutton C (2010). Joel Monaghan in tears after quitting the Raiders. *Sydney Morning Herald*, 9 November. [Online] Available: www.smh.com.au/rugby-league/league-news/joel-monaghan-in-tears-after-quitting-the-raiders-20101109-17lff.html [Accessed 14 February 2013].

Beirne P (2000). Rethinking bestiality: towards a concept of interspecies sexual assault. In A Podberscek, E Paul & JJ Serpell (eds), *Companion animals and us: exploring relationships between people and pets* (pp313–31). Cambridge: Cambridge University Press.

Bernardi cancels London speech after bestiality remarks (2012). *ABC News*, 22 September. [Online] Available: www.abc.net.au/news/2012-09-22/bernardi-pulls-out-of-uk-speaking-engagement/4275422 [Accessed 22 September 2012].

Bolliger G & Goetschel A (2009). Sexual relations with animals (zoophilia): an unrecognized problem in animal welfare legislation. In A Beetz & A Podberscek (eds), *Bestiality and zoophilia: sexual relations with animals* (pp23–48). New York: Berg.

Dekkers M (1994). *Dearest pet: on bestiality*. London: Verso.

Doyle J (2010). ACT to again criminalise sex with animals. *ABC News*, 9 December. [Online] Available: www.abc.net.au/news/2010-12-09/act-to-again-criminalise-sex-with-animals/2368470 [Accessed 22 April 2012].

Dutton C (2010). Ultimatum to raiders: sack Monaghan or risk sponsors. *The Canberra Times*, 6 November, pp1–2.

Franklin A (1999). *Animals and modern cultures: sociology of human–animal relations in modernity*. London: Sage.

Friedman A (2012). My wife! My best friend! [Cartoon] *The New Yorker*, 24 September. [Online] Available: www.newyorker.tumblr.com/post/31814486301/cartoon-of-the-day-by-andy-friedman-for-more-from [Accessed 24 September 2012].

Fudge E (2008). *Pets*. Stocksfield, UK: Acumen.

Gavel T (2010). The Monaghan controversy: how quickly things go pear-shaped. *CityNews*, 18 November, pp14.

Goldsworthy P (1995). *Wish*. Sydney: Angus & Robertson.

Levy N (2003). What (if anything) is wrong with bestiality? *Journal of Social Philosophy*, 34(3): 444–56. DOI: 10.111/1467_9833.00193.

Mascord S (2011). Interview with Joel Monaghan. *Rugby League Week*. 16 February, p17.

Miletski H (2009). A history of bestiality. In A Beetz & A Podberscek (eds), *Bestiality and zoophilia: sexual relations with animals* (pp1–22). New York: Berg.

Moloney J-P & Dutton C (2010). Monaghan loyal to the end. *The Canberra Times*, 1 November, p1.

Morriss P (1997). Blurred boundaries. *Inquiry: An Interdisciplinary Journal of Philosophy*, 40(3): 259–89. [Online] Available: www.dx.doi.org/10.1080/00201749708602452 [Accessed 22 April 2013].

Moses A (2010). A dog act: Twitter leaker breaks cover. *WAtoday*, 5 November. [Online] Available: www.watoday.com.au/technology/technology-news/a-dog-act-twitter-leaker-breaks-cover-20101105-17gex.html [Accessed 28 September 2012].

Packham B & Kerr C (2012). Senator Cory Bernardi quits 'after one mistake too many'. *The Australian*, 19 September. [Online] Available: www.theaustralian.com.au/national-affairs/cory-bernardi-quits-over-bestiality-comments/story-fn59niix-1226477232022 [Accessed 19 September 2012].

Rake (2010). R vs Chandler. *Rake*, series 1, episode 5. Written by P Duncan. Sydney: ABC TV.

RSPCA (n.d.). Five freedoms for animals. RSPCA. [Online] Available: www.kb.rspca.org.au/Five-freedoms-for-animals_318.html [Accessed 22 September 2012].

Ryle J (1994). Yearning's outer limits: 'Dearest pet: on bestiality' [Book review]. *The Independent*, 17 July. [Online] Available: www.independent.co.uk/arts-entertainment/book-review--yearnings-outer-limits-dearest-pet-on-bestiality--midas-dekkers-verso-1895-pounds-1414470.html [Accessed 24 September 2012].

Singer P (2001). Heavy petting. *Nerve*. [Online] Available: www.utilitarian.net/singer/by/2001----.htm [Accessed 18 September 2013].

Tiffin H (2009). Animal writes: ethics, experiment and Peter Goldsworthy's *Wish*. *Southerly*, 69: 36–56.

Toffoletti K (2007). How is gender-based violence covered in the sporting news? An account of the Australian Football League sex scandal. *Women's Studies International Forum*, 30(5): 427–38.

Vollum S, Buffington-Vollum J & Longmire DR (2004). Moral disengagement and attitudes about violence and animals, *Society & Animals,* 12(3): 209–35. [Online] Available: http://libra.msra.cn/Publication/44428587/moral-disengagement-and-attitudes-about-violence-toward-animals [Accessed 23 April 2013].

Wilson R (2010). With mates like these. *The Daily Telegraph*, 6 November. [Online] Available: www.heraldsun.com.au/sport/nrl/with-mates-like-these/story-e6frfgh6-1225948534573 [Accessed 22 June 2011].

5

Encounters with Antarctic animals in ABC's *Catalyst*

Sophie Fern, Kate Nash and Elizabeth Leane

The animals that inhabit the Antarctic continent and its surrounding waters have a unique relationship with humanity. They are the only creatures to occupy a continent with no indigenous human population or permanent human inhabitants. While people venturing into far southern waters – whalers, sealers and explorers – have interacted directly with Antarctic marine mammals and birds for centuries (to these creatures' significant detriment),[1] the wider public has primarily encountered them through zoos (Martin 2009)[2] or through the eyes of others in drawings, paintings, specimens and, eventually, still and moving images. Although human contact with the region has increased in recent years with the expansion of scientific presence and the development of large-scale cruise-ship tourism, the number of people who have immediate experience of the continent – and hence of its wildlife in situ – is very low. While an 'unmediated' encounter with animals is difficult to imagine, it is nonetheless the case that human perceptions

---

1 Antarctic marine species were hunted, some almost to extinction, for their pelts (fur seals) or for oil (whales and penguins). See Commission for the Conservation of Marine Living Resources. Retrieved on 23 October 2012 from www.ccamlr.org/en/organisation/history.
2 While sub-Antarctic animals have been kept in zoos since the mid-19th century, Antarctic animals such as emperor penguins are harder to house, and even now feature in relatively few zoos and aquaria. King penguins first arrived at the Zoological Garden in Regent's Park, London, in 1865 (Martin 2009, 77–78).

of Antarctic animals, like the icescape itself, are especially dependent upon (primarily visual) texts.

Interrogating the visual representation of Antarctic animals draws attention to the ways in which public understanding is shaped. The concept of charisma provides a framework for considering how audiovisual representations underpin specific human–animal relationships, which can in turn form the basis of conservation efforts. Penguins, for example, a species with established high appeal to humans (Woods 2000), have acted as a 'flagship' for conservation efforts in the region (Stokes 2006, 362), often functioning as a synecdoche for the Antarctic wilderness and its ecosystems. With the far south the subject of increasing attention in the context of the climate-change debate, this role continues to be significant. Research has shown that animal charisma has both biological and social determinants. Thus while there are some species – such as penguins – that seem to generate an innate cute response in humans,[3] it is nonetheless possible to deliberately generate human sympathy for Antarctic animals who do not produce this automatic reaction.[4]

This chapter examines the representations of two very different kinds of Antarctic animals – penguins and invertebrates – in the contemporary Australian science journalism program *Catalyst*. This series, we argue, deploys diverse kinds of charisma to create environmental concern for a range of Antarctic species. Our analysis shows that not only does *Catalyst* draw on diverse kinds of charisma, it does so even when presenting the same species. This has important implications for conservation: charisma has been recognised as being 'vital for the enrolment of public support for conservation' (Lorimer 2007, 926). By conveying a variety of charismatic properties in animals, rather than limiting themselves to exploiting the 'cute response,' television journalists can maximise audience sympathy for endangered animals.

---

3   This fairly self-evident term can be defined as 'the feeling people experience on encountering beings whose appearance stimulates a desire to hold, cuddle, and protect them' (Milton 2011, 68).

4   It is not easy to separate a physiological charismatic response from responses that have been socially produced. As we discuss below, penguins were not necessarily considered 'cute' by Europeans on first encounter, although they were very readily anthropomorphised. See Martin (2009) for a survey of changing human responses to the animals.

Antarctic wildlife on screen

If human access to Antarctic animals is largely dependent on texts, these texts themselves are constrained in turn by the logistical and financial challenges of filming in high southern latitudes. At the turn of the 20th century, when land-based exploration of the continent began, expeditions provided key sources of images of Antarctic animals in the form of photographs and documentary films in which wildlife was frequently the lead attraction (Leane & Nicol 2013). Documentaries have continued to be the key screen genre produced about the continent, with the financial outlay of sending a film crew to Antarctica meaning that most productions tend toward the blue chip[5] – David Attenborough's *Life in the freezer* is one example; the bestselling documentary *March of the penguins* is another.

In the last couple of decades, however, tourist ships and arts residencies within national research programs have seen an expansion in the diversity of screen production in the continent. In particular, it is now not unusual for magazine-style popular television series to feature one or more segments set in the continent. From Australia, for example, comedian Andrew Denton travelled as resident artist with the Australian Antarctic Division (AAD) in 1993–94 to film an episode of his comedy series *The money or the gun*, leaving a camera with base personnel over winter for extra footage. In 2006, the AAD invited a team from the commercial television children's series *Totally wild* – which focuses on adventure, wildlife and environment – to the continent to film six episodes. In 2009 Mark Horstman, a reporter with *Catalyst*, travelled to East Antarctica with the AAD, while his colleague Paul Willis went to the other side of the continent, the Antarctic Peninsula, as a scientific advisor on a cruise ship.

While the standard format of the blue-chip wildlife film is well established – the focus on animal behaviour and spectacular scenery, the portrayal of the natural world as untouched by humanity – the approach of the magazine science program is more complex and less discussed. To understand *Catalyst*'s depiction of Antarctic animals and their ecosystems, it is necessary to understand both the concept of

---

5   Blue-chip natural history describes a genre that usually depicts spectacular wildlife in natural settings with an omniscient narrator (see Bousé 2000, 14–15).

nonhuman charisma and the genre of the magazine popular science program in more detail.

Nonhuman charisma

The term 'charisma' – as in 'charismatic megafauna' or 'charismatic species' – is frequently used in social-science and humanities scholarship focused on wildlife conservation, particularly in relation to the selection of 'flagship' species that act as catalysts for the protection of a wider ecosystem (Verissimo et al. 2012; Home et al. 2009; Sergio et al. 2006). The meaning of 'charisma' is often assumed in these publications. Lorimer (2007) is the first to define the term in detail and to categorise different forms of charisma. He describes it as 'the distinguishing properties of a nonhuman entity or process that determine its perception by humans and its subsequent evaluation' (915). Importantly, this quality is 'subject to anthropogenic manipulation' (915) – something that has been explored by other researchers looking at particular case studies (eg Milton 2011, 67–77). Lorimer (2007) outlines a 'three part typology' of nonhuman charisma – ecological, aesthetic and corporeal – that provides a useful framework through which to understand *Catalyst*'s manipulation of its audience's sympathies for threatened Antarctic wildlife.

Ecological charisma is endowed on species which humans find comparatively easy to detect and relate to (Lorimer 2007, 916). Humans find certain species more detectable and familiar, depending on a variety of parameters including 'size, colour, shape, speed and degree of movement,' as well as the 'nature and frequency of any human–nonhuman encounter' (917). Lorimer (2007, 918) notes by way of negative example that 'the inaccessible benthic ecology and microscopic and indistinct anatomy of deep sea nematodes . . . render their conservation much more difficult' – they lack ecological charisma. Penguins, as bipedal creatures, are easily detectable by and 'relatable' (so to speak) to humans, therefore they are high in ecological charisma. Seals, as relatively large mammals that spend significant time on land, also have their share of the quality, but marine invertebrates, inaccessible and lacking in familiar features or behaviours, score very low on this scale.

The second and third kinds of charisma that Lorimer (2007, 918) outlines are grouped together as types of 'affective charisma', a form which relates to the 'emotions, affections, and motivations triggered by organic nonhumans'. 'Corporeal charisma', generated by 'corporeal interactions with an organism in the field' (918), has comparatively little relevance to Antarctic animals, except inasmuch as they are a tourist attraction; the overwhelming majority of people never interact with Antarctic animals in the field. Far more significant is 'aesthetic charisma', which relates to the aesthetic properties of an organism's appearance and behaviour when encountered visually by an observer either in the flesh or as a textual inscription (918).

Importantly, the affect generated by aesthetic charisma need not be positive. While some humans are attracted to 'cuddly' charisma – the cute response generated by the anthropomorphic preference for creatures with recognisable facial characteristics, and particularly neotenic features[6] – others respond to its polar opposite – the negative aesthetic of 'feral' charisma (920). This 'yuck factor' is generated by 'organisms such as insects, that are radically different to anthropocentric norms' (920). It is exotic otherness, rather than cosy familiarity, that appeals to humans in these species. While 'cuddly' charisma generally trumps 'feral' charisma, the latter nonetheless has its advocates, and relies on 'a sense of respect for the other and for its complexity, autonomy, and wildness' (920).

In the case of the two extremes of Antarctic species – penguins and marine invertebrates – it would seem obvious to exploit ecological and aesthetic 'cuddly' charisma in the first case, and where possible 'feral' charisma in the second. However, a close examination of the *Catalyst* episodes at issue here shows that viewers are manipulated, in both cases, to appreciate both kinds of charisma.

About *Catalyst*

Catalyst is a weekly science journalism television program produced by the Australian Broadcasting Corporation (ABC). The ABC is a com-

---

6 Neoteny is the retention in some species of juvenile characteristics into adulthood.

prehensive non-commercial national broadcaster whose culture is reflective of principles of public service broadcasting (PSB), particularly the provision of quality content that is not driven by commercial imperatives (Debrett 2009, 807). An important PSB value and marker of program quality is an educative agenda (Hawkins 1999). Reflecting this educative agenda, science programming has long been a feature of the television schedule with a science unit established in the 1970s (Bowden & Borchers 2006, 148). In the mid-1980s the half-hour magazine style format was established with *Towards 2000* and later *Quantum* and most recently *Catalyst*. Each episode consists of three to five journalistic stories that focus both on the scientific process and social context. This approach can be seen as an attempt to educate the citizen-viewer to enable him/her to participate in scientifically informed decision-making (Dhingera 2006, 100). In the 1990s ABC Science faced increasing pressure to balance these educative and journalistic functions with the need to be an 'entertaining prime-time show' (Gilling 1986, 42). *Catalyst*, which began in 2001, continues to blend information, journalism and entertainment in short digestible segments.

Science programs and wildlife television are specialised in their subject matter and both are primarily driven by the principles of journalism and television broadcasting rather than of science (Lehmkuhl et al. 2012). Situating *Catalyst* with reference to science programming in general is difficult since it seems to display a blend of characteristics. Lehmkuhl et al. (2012) identify five broad types of science programs:

1. informational programs which predominantly consist of science news broadcasts
2. popularisation programs characterised by an informative stance and generally consisting of interviews connected by narration
3. entertainment programs which aim to both educate and entertain, driven by celebrity presenters, with science often a secondary consideration
4. advice programs that focus on health and environmental issues and which seek to give the audience practical information
5. advocacy programs which address scientific issues but whose main focus is on fields other than science, most often politics.

While *Catalyst* most closely corresponds to the popularisation form, its journalistic approach aligns it to some extent with advocacy programs (particularly in the context of environmental issues) and, in its short presenter-led segments, choice of subjects and style, it is unquestionably entertainment-focused.

Catalyst in Antarctica

In the summer of 2009 Horstman and Willis travelled to Antarctica to produce several stories for *Catalyst*. Over the next two years *Catalyst* broadcast eight stories focused on the Antarctic region, seven of which dealt with animals that live there: 'Penguin wave' (22 September 2011), 'Penguin DNA' (26 May 2011), 'Iron whales' (14 April 2011), 'Southern Ocean sentinel' (29 April 2010) 'Trawl team' (20 May 2010), 'Penguins: winners and losers' (26 August 2010) and 'Mysteries of the emperors' (2 September 2010). Five groups of animals are represented in the programs: whales, seals, fish and invertebrates, penguins, and other sea birds.

In spite of the diversity suggested by this list, two groups – the seabirds and seals – are represented only within short montage sequences at the beginning or end of the story. Two other groups – the penguins and invertebrates – are the most visible, with penguins featuring in every story, and two stories – 'Southern Ocean sentinel' and 'Trawl team' – focusing particularly on invertebrates. While on one level the prominence of invertebrates is surprising given the tendency of blue-chip wildlife programs to focus on charismatic megafauna (Bousé 2000), their presence can be explained in terms of the scientific agenda of the program. In the following section, we analyse the different techniques used to frame the animals and to activate their potential charisma. While it might seem that penguins and invertebrates are at opposite ends of the charisma spectrum, we suggest that the situation is far more complex.

Penguins

Unremarkably for animals with high ecological and 'cuddly' charisma, penguins feature heavily in the stories produced by *Catalyst*. They are the principal subject of four stories but are also featured in the land-scape in stories where they are not scientifically relevant (such as in 'Trawl team'). As part of the Antarctic environment, penguins can provide a focus for attention, act as a reference for the scale of image, or serve a metaphoric role. Their comic potential is often exploited, with penguins providing contrast to the serious business of scientific re-search. In 'Southern Ocean sentinel',[7] for example, a sequence of shots in which scientists take ice samples is rendered mildly amusing through the intercutting of several shots of penguins (which may have been taken at a different time or even location). Drawing on audience fa-miliarity with continuity editing, the scene implies that the penguins are 'watching' the scientists at work. The addition of chattering penguin sounds enhances the comic effect. The frequent use of chattering, laughter-like sound (again, almost always likely to be added in post-production) serves as an aural signifier of the penguin. Significantly, the use of sound can be considered anthropomorphic, drawing attention to the fact that penguins, like humans, are vocal communicators.

Two of the four *Catalyst* stories focused explicitly on penguins, and looked at conservation issues. 'Penguins: winners and losers' and 'Mys-teries of the emperors' consider the vulnerability of penguin species in the face of climate change. In these stories image and narration fre-quently combine to both anthropomorphise the penguin and exploit the animals' comic value. 'Penguins: winners and losers' considers the impact of climate change on two penguin species, the Adélie and the gentoo. While the relatively adaptable gentoo penguin has so far coped well with changes in sea ice on the Antarctic Peninsula, the Adélie pen-guin population seems to be declining rapidly. The story is introduced by reporter Paul Willis who is standing in the middle of a penguin colony and directly addressing the viewer in a piece to camera. This fre-quently used technique serves to suggest the tameness of the penguin and the potential for human–penguin interaction. The two penguin species are introduced to the viewer with extended close-ups that, in combination with the narration, teach the audience how to tell the two species apart.

---

7   The sequence can be found approximately two minutes and 50 seconds into the story.

The Adélie penguin is introduced with a single shot tilting up to focus on the face. The gentoo are introduced with a slightly comic sequence in which one bird, making a characteristic laugh-like call, is rushed at by a second that appears to try to tackle the first. The contrast between the fate of the Adélie penguins and the fate of the gentoos is rendered visually with shots of sprawling, raucous gentoo colonies intercut with small groups of Adélies – often seemingly isolated on drifting ice islands. An interview with biologist Dr Louise Emmerson, in which she explains that the vulnerability of the Adélie penguin is due to a lack of flexibility in breeding and foraging, provides an opportunity to show the Adélies feeding chicks. In spite of the story's focus on the vulnerability of the Adélie penguins, ultimately it suggests that both species are likely to suffer from the rapid changes facing the Antarctic ecosystem.

The threat to the survival of a rare species is also a theme of the story 'Mysteries of the emperor'. Here the research of Dr Barbara Wienecke into the diving and foraging habits of emperor fledglings provides the narrative focus with a broader reference to 'unlocking the secrets to their survival'. The opening depicts the emperor penguin as a hardy species with a sequence of a huddle of brooding males in the midst of a snowstorm. A reference to a famous 1911 journey during the Scott expedition to collect an emperor egg provides an opportunity for reenactment, accompanied by an extract from Apsley Cherry-Garrard's famous account in *The worst journey in the world*: 'Take it all in all, I do not believe anybody on Earth has a worse time than an Emperor Penguin' (Cherry-Garrard 1922, xvii). A century later, we are told, the emperor penguin 'remains a majestic mystery to modern science'.

While the emperor is 'hardy', 'majestic' and 'exquisite', it is also framed as human-like in many of its characteristics, particularly its familial relations. All penguins invite a degree of anthropomorphism by virtue of their bipedal gait, but it is the largest species, the emperors, in which this is most noticeable. In 'Mystery of the emperor', the research serves as a pretext for a narrative that focuses on intergenerational relationships, providing opportunities for anthropomorphism in the depiction of parent–chick relationships (the focus of the highly successful documentary *March of the penguins*). After the research project is introduced through footage of Dr Wienecke studying a breeding colony at Amanda Bay, the viewer is taken into the heart of the colony. The first shot is a spectacular close-up of a single chick, shot from a low angle so that the chick appears to be looking down into the camera. As the chick chirps and nods its head, two adult birds seem to be standing alongside: one immediately next to the chick and the other behind and slightly to one side. A sequence of parent/chick shots follows, which includes a father carrying a chick on his feet, a large group of chicks, and a shot that follows a chick as it wanders through the colony. The scientific research is introduced with an intertextual reference to the animated movie *Happy Feet* that connects the numerous shots of chicks to the filmic narrative of triumph against the odds. The research will reveal 'where Happy Feet gets to when he or she leaves home'. Several shots of single adult birds serve to suggest that the chicks' departure is experienced as an emotionally charged time.

The research requires the catching and radio tagging of several fledglings. In the description of 'how to catch an emperor penguin' there is a confluence of comic and anthropomorphic elements. Music sets a comic mood, an oft-used device in establishing a humorous frame in relation to penguins. Dr Wienecke explains that 'you watch your group from a safe distance very carefully and you pick your candidate'. A shot of a large group of chicks is intercut with an individual chick stretching and flapping its wings. Dr Wienecke then goes on to explain that the scientist must pretend to be interested in every bird but the chosen one because 'Somehow, they get a sense of "oh my god, she's after me", and they will just turn around and it's extraordinary how fast they can move on the ice'. A group shot at this point seems to depict birds looking at the ground, apparently attempting not to be noticed by the researcher. This is followed by a shot of a single chick apparently 'running away' through the group. In this short sequence the anthropomorphic and comic come together to suggest a connection between penguin and human both in terms of characteristics and humour.

Having established an empathy between human and bird, the story shifts to consider the impact of environmental change on the emperor population. A sequence of shots from 1994, 'a bad breeding year', shows Dr Wienecke collecting and studying the bodies of dozens of emperor chicks. As she talks about the challenges facing the emperors, there is a shot of a single bird apparently struggling to get itself back onto the ice. Following a narration that focuses on the scientific mystery of the species, Dr Wienecke closes the story by linking human and emperor in a much more troubled relationship: 'It'll be devastating if . . . [we] find that we have actually caused the extermination of a magnificent species like the emperor penguin'.

Two techniques – anthropomorphism and comedy – are thus exploited in these episodes to engender viewer sympathy for the penguins in their threatened environment. While the ease with which penguins can be represented anthropomorphically relates directly to their ecological charisma, their construction as a comic subject is more complex. An analysis of recent penguin films argues that while their anthropomorphism relates to their 'perceived likeness to humans: as flightless birds that stand upright, with arm-like flippers', at the same time 'as fish-like birds confined largely to remote Southern Hemisphere locales, penguins act as exotic others to humanity. This combination of strangeness and familiarity is key to their wide appeal and flexible cultural significance' (Leane & Pfennigwerth 2011, 30–31). One explanation of the frequent recourse to the comic in representations of penguins is this incongruity: they evoke a mixture of the familiar and the strange, or, to use Lorimer's terms, of ecological and 'feral' charisma. While the latter may not seem intuitively obvious – penguins do not characteristically evoke a 'yuck factor' – an examination of the history of human–penguin encounters indicates that penguins were initially perceived by Europeans as something quite alien and 'other' (Martin 2009, 40). Increasing exposure to penguins through films, photographs and zoo exhibits has decreased this sense of strangeness, which nonetheless remains underlying the comic effect produced by the creature – a figure like a tuxedoed man that waddles on ice like a toddler and (although a bird) swims in the sea.

The exploitation of this combination of ecological, aesthetic and 'feral' charisma in the representation of penguins is not new to *Catalyst*; very similar conventions appeared a hundred years earlier in the expedition films of Herbert Ponting and Frank Hurley.[8] More innovative, however, is *Catalyst*'s combination of these forms of charisma in their construction of Antarctic marine invertebrates.

---

8   See Leane and Nicol (2013) for an analysis of the representation of animals in these films, in which the exploitation of penguin's comic value through accompanying music and intertitles or voiceovers is apparent.

Invertebrates

Insects, according to Lorimer, generate 'popular feelings of antipathy and distrust' in humans as terrestrial invertebrates, due to 'their radial alterity to humans in terms of size, ecology, physiology, aesthetics and modes of social organization' (Lorimer 2007, 920). Marine invertebrates – more remote from humans and similarly lacking in aesthetic charisma – are also likely to produce the same negative response, as Lorimer's comment about deep-sea nematodes (quoted above) suggests. Unsurprisingly, they do not often feature in popular culture (Leane & Nicol 2011).[9] It is significant, then, that two of the six *Catalyst* episodes examined here, 'Trawl team' and 'Southern Ocean sentinel', give a prominent role to these organisms.

The only hope for those trying to engender audience sympathy for marine invertebrates would seem to be 'feral' charisma, and the opening of 'Trawl team' clearly takes this approach, framing the deep-sea environment as an alien world. Against images of pink skies, the script talks about a 'hidden world' cloaked in ice. More is known, the narration tells us, about the surface of the moon than the deep seas of the Southern Ocean. In a piece to camera, reporter Mark Horsham invites the spectator to consider the research vessel *Aurora Australis* as a spaceship 'flying high above an unexplored planet for the very first time'. The only way to get a glimpse of the 'vast alien world far below' is by lowering a net to capture images, and samples when conditions permit. From the fragments brought up from below, scientists attempt to piece together a picture of life on the ocean floor. A rapidly edited sequence of still images of marine invertebrates gives the viewer a sense of the diversity of life below the surface, but the quick cutting between shots works to reinforce the alien frame by denying the viewer the opportunity to make sense of the images. The dozen images flash before the eyes in a confusing rush suggesting a bizarre world unlike anything with which the viewer might be familiar.

As the camera plunges beneath the surface the narration emphasises the hostility of the deep-sea environment with reference to 'sub-zero temperatures, crushing pressures and utter darkness'. Yet this im-

---

9   The film *Happy Feet two*, which features two krill voiced by Brad Pitt and Matt Damon, is a notable exception.

age of a hostile and foreign environment is accompanied by depictions of 'a rugged realm of delicate beauty'. The footage from the underwater camera shows views of a pastel-coloured faunal assemblage that look like a muted version of a coral reef, and the voice-over uses these terms. This reef is delicate and fragile, and the views of it are fleeting. The camera lingers on two shots of a feathery-looking creature – an unstalked crinoid – gliding gracefully through the water. The effect is to establish the benthic environment and its animals as radically other but also graceful and beautiful. 'Feral' charisma combines with another form of aesthetic charisma that is not 'cuddly' or familiar, but nevertheless creates a positive aesthetic response – a similar response, one might guess, to that produced by inorganic objects that are considered beautiful.

Not only are these animals beautiful, they are an integral part of the ocean ecosystem. One of the research team makes the claim that, 'without the wonderful creatures associated with the benthic habitats, our cycle of life throughout the ocean could possibly cease to exist'. This ecological frame sets up the political problem that the science will ultimately address: the threat of overfishing. Significantly, the first pieces of evidence uncovered by the underwater cameras indicate illegal fishing activity. A vast empty underwater landscape contrasts markedly with the abundance of earlier underwater shots. In the midst of the desolate landscape, the team discovers regularly spaced parallel furrows, evidence of long-line fishing. The narration puts the discovery in its political context, stating that it is a 'reminder of why we're here'.

Eventually the team hit a 'biological hot spot' with the trawl nets disgorging mountains of swarming benthic creatures onto the trawl deck. There is a sense of excitement and discovery as the animals are whisked into the wet lab to be sorted and thereby rendered scientifically meaningful. With the creatures removed from their environment, increasingly unresponsive and slowly dying, it becomes a task of the presenter to animate and describe the various species. In a close shot a giant octopus is transferred from one container to another, its sheer size signifying the abundance of the marine environment. Three smaller octopus-like animals are filmed from below the glass container in which they have been placed, providing an opportunity to emphasise their unusual body shape. A gaping fish with an enormous head is held up for the camera as the narration informs the audience that almost all of these benthic groups 'have more than half of their species endemic to the Southern Ocean – that is, they only live here'. A series of still images of individual animals taken against a black backdrop affords the audience a closer look at the diverse collection of specimens. This is, it seems, another attempt to go beyond 'feral' charisma – by visually presenting these invertebrates as accessible individuals, the camera endows upon them some measure of ecological charisma.

Throughout the story we are encouraged to see deep-sea fauna as alien and bizarre, yet also beautiful, diverse (in the end 472 new species were found) and ecologically significant. In its conclusion, the story highlights the persistent danger to the ocean environment as the trawl team catches an abandoned long-line. In the closing narration, accompanied by a sequence of shots of iconic Antarctic animals such as seabirds, seals and whales, the narration emphasises the beauty and significance of these animals, linking them to the benthic creatures uncovered by the trawl team within a clear message of conservation: 'we ended up discovering a highly productive part of the ocean that no-one knew existed and deserves protection'. In the end then, the frame returns to the charismatic megafauna but not until several kinds of charisma – ecological, feral, and something more purely aesthetic – have been deployed to focus viewers positively on invertebrates themselves.

The second episode that concentrates on invertebrates, 'Southern Ocean sentinel', deals with pteropods, microscopically small molluscs that have a spiral snail shell, and a foot that has evolved into wing-like lobes that protrude like overgrown ears from the bottom of the shell. The story examines the possible consequences to marine fauna of lowered pH levels (that is, more acidic conditions) in the ocean. Although pteropods are otherworldly and removed from that with which the viewer is familiar, they are not feral. There is nothing 'yucky' about them – where insects and other anthropods generate fear partly because they are recognisably animal while nonetheless profoundly different from humans, pteropods are too radically dissimilar from humans to evoke even this form of (potentially pleasurable) discomfort. It is unclear whether the viewers will actually perceive pteropods as animals as they have so few of the features that are associated with the stereotypical animal.

The segment must therefore establish pteropods as worthy of the viewers' attention in some other way. As in 'Trawl team', this is achieved by drawing on an aesthetic charisma more basic than the 'feral' or the 'cuddly'. Using close-up, enormously magnified photographs of live pteropods on a black background, the episode foregrounds the beauty of the creatures.[10] These creatures are presented as beautiful, near-inorganic objects, like jewels, and the shots of them are all static. The piece, however, carefully emphasises the effects of acid on the pteropods' shells, which are the only part of the organism with which the viewer will be familiar as they are microscopically sized whorls that look just like those of land snails. Because the viewer understands that land snails cannot survive outside of their shell, they can also understand that pteropods would die without their shells. Here, an ecological charisma[11] is metaphorically borrowed from pteropods' more familiar terrestrial counterparts.

While only these two stories take invertebrates as their primary focus, several others ('Iron whales', 'Penguins: winners and losers', 'Southern Ocean sentinel') deal with krill as a key organism in the Southern Ocean food chain. Whereas previous research has emphasised the de-individualisation of krill in popular representations – they are reduced almost to vegetation, as 'forage' for other animals (Leane & Nicol 2011, 138) – *Catalyst* is significant in making an attempt to visually endow the species with a charisma of sorts. This is most obvious in the extended visual representation of krill in the story 'Iron whales'.

The story opens with a spectacular shot taken from the air of a baleen whale gulping 'mountains' of krill. The question to be explored is whether whale faeces might provide much-needed iron to support the growth of phytoplankton, microscopic plants which effectively remove carbon from the atmosphere. While the first half of the story focuses on the whales and the challenges of studying their faeces, it then goes on to look at krill, 'the perfect mobile concentrators of iron'. Krill are initially

---

10   By magnifying an animal that would otherwise be all but invisible, the production is following a familiar documentary practice of expanding what Keith Beattie calls 'the realm of the visible' (Beattie 2008). Micro-cinematography techniques like these have been used since the early work of practitioners like Jean Painleve, and they allow the viewer to tap into their desires to see that which is normally hidden. For a review of the work of Jean Painleve, see Masaki Bellows, McDougall & Berg (2000).
11   In a survey conducted in Northeast India drawing on 255 residents and 105 tourists, snails were amongst the ten most liked terrestrial invertebrates (Barua et al. 2012).

introduced as swarms that graze on the phytoplankton. This verbal reference to swarming is accompanied by a wide shot of a writhing cloud of krill. While this insect-like depiction suggests a 'feral' appeal – the otherworldliness of the swarm – the next shot depicts a single krill in close-up, emphasising its swimming action. Similarly, a shot of a small group of krill allows for a focus on individual animals and, although the movement of the krill here seems more random by virtue of the increase in numbers, the motility of individual krill is evident.

The tension between representing krill as an insect-like swarm or mass and as individual active animals continues in the next sequence. In order to represent the scientific process of capturing krill there is a sequence depicting krill harvesting. Here the krill are inert; stored in buckets, they are less individualised animals than a pinky substance to be decanted from one container to another as a scientific object. However, a cut back to some close-up shots of the krill immediately follows. In particular a long (three-second) close-up of an individual krill, shot against black (suggesting a kind of transparent glow) shows the animal swimming as though toward the camera, its 'face' clearly visible – an attempt, perhaps, to engender a sense of affective, if not convincingly 'cuddly', charisma. Following this moment of visual connection with the krill, however, we return to the ocean and the krill once again become a swarm of soon-to-be nutrients.

Various kinds of charisma thus oscillate in the visual framing of this marine invertebrate species. The filmmakers deliberately activate different forms of audience sympathy: they deploy a readily available 'feral' charisma, but also encourage recognition of ecological charisma by presenting marine invertebrates as individuals and making links with more familiar terrestrial animals. In addition, they aestheticise the animals by including shots that emphasise a visual appeal quite different from either 'cuddly' or 'feral' charisma, ironically generating a form of connection by constructing the animal as beautiful object. While these techniques are in some ways contradictory, they show a willingness to go beyond the cute response to find ways of encouraging humans to see a connection with animals who might initially be perceived as insurmountably 'other'.

Conclusion

Catalyst deploys a variety of forms of nonhuman charisma in its de-piction of Antarctic species, whether these are flagship species (such as penguins) or the marine invertebrates that are, traditionally, anything but flagships. This combination of approaches, particularly in regard to invertebrates, can be tied to *Catalyst*'s remit to entertain and instruct. Where ecological and 'cuddly' aesthetic charisma speaks to the instinct-ively popular, an appreciation of 'feral' charisma seems to be something learned with time and experience: Lorimer (2007, 920) suggests, in the case of insects, it is writers and conservationists – particularly entomo-logists – who are particularly likely to prefer 'feral' to 'cuddly' charisma. In its deployment of a range of types charisma to engender sympathy for Antarctic animals under environmental threat, *Catalyst* is not only instructing its viewers in science, it is teaching them new and diverse ways to acknowledge the appeal of other creatures.

Works cited

Barua M, Gurdak DJ, Amed RA & Tamuly J (2012). Selecting flagships for invertebrate conservation. *Biodiversity Conservation,* 21: 1457–76.

Bousé D (2000). *Wildlife films.* Philadelphia: University of Pennsylvania Press.

Bowden T & Borchers W (2006). *50 years: Aunty's Jubilee! Celebrating 50 years of ABC Television.* Sydney: ABC Books.

Catalyst (2010–11). Australian Broadcasting Corporation. [Online] Available: www.abc.net.au/catalyst/stories/by-date/2012/ [Accessed 12 December 2012].

Cherry-Garrard A (1922). *The worst journey in the world: Antarctic 1910–1913.* London: Constable & Co.

Debrett M (2009). Riding the wave: public service television in the multi-platform era. *Media Culture and Society,* 31(5): 807–27.

Life in the freezer (1993). Produced by A Fothergill, presented by D Attenborough. BBC.

Gilling D (1986). The quantum view. *Media International Australia,* 42: 54.

Happy Feet two (2011). Directed by G Miller. Warner Bros Pictures.

Hawkins G (1999). Public service broadcasting in Australia: value and difference. In A Calabrese & J Burgelman (eds), *Communication, citizenship and social policy* (pp173–87). Oxford: Rowman & Littlefield.

Home R, Keller C, Nagel P, Bauer N & Hunziker M (2009). Selection criteria for flagship species by conservation organizations. *Environmental Conservation*, 36(2): 139–48.

Leane E & Nicol S (2013). Filming the frozen south: animals in early Antarctic exploration films. In A Pick & G Narraway (eds), *Screening nature: cinema beyond the human* (pp127–42). Oxford: Berghahn Books.

Leane E & Nicol S (2011). Charismatic krill? Size and conservation in the ocean. *Anthrozoös*, 24(3): 135–46.

Leane E & Pfennigwerth S (2011). Marching on thin ice: the politics of penguin films. In C Freeman, E Leane & Y Watt (eds), *Considering animals: contemporary studies in human–animal relations* (pp29–40). Farnham, UK & Burlington, US: Ashgate Publishing.

Lehmkuhl M, Karamanidou C, Mörä T, Petkova K & Trench B (2012). Scheduling science on television: a comparative analysis of the representations of science in 11 European countries. *Public Understanding of Science*, 21(8): 1002–18.

Lorimer J (2007). Nonhuman charisma. *Environment and Planning D: Society and Space*, 25: 911–32.

March of the penguins (2005). Directed by L Jacquet. Warner Bros Pictures.

Martin S (2009). *Penguin*. London: Reaktion Books.

Milton K (2011). Possum magic, possum menace: wildlife control and the demonisation of cuteness. In Freeman C, Leane E & Watt Y (eds), *Considering animals: contemporary studies in human–animal relations* (pp67–77). Farnham, UK & Burlington, US: Ashgate Publishing.

Sergio F, Newton I, Marchesi L & Pedrini P (2006). Ecologically justified charisma: preservation of top predators delivers biodiversity conservation. *Journal of Applied Ecology*, 43:1049–55.

Stokes D (2006). Things we like: human preferences among similar organisms and implications for conservation. *Human Ecology*, 35: 361–69.

Verissimo D, Barua M, Jepson P, MacMillan DC & Smith RJ (2012). Selecting marine invertebrate flagship species; widening the net. *Biological Conservation*, 145: 4.

Woods B (2000). Beauty and the beast: preferences for animals in Australia. *The Journal of Tourism Studies*, 11 (2): 25–35.

Part II
History, art and literature

6

The donkey and Mr Simpson: remembering the donkey in the Anzac legend

Jill Bough

> The ass is already immortal, for the Salvager of Souls used him. Let us again use the ass as symbolical along with Simpson of the salvaging of human wreckage in war.
>
> *The Argus*, October 1933

> Donkeys operate as emotionally affective, spiritual and patriotic media and subjects of commemoration.
>
> Williams 2011

Ken Inglis, in his fascinating exploration of war memorials in Australia, starts his book with his experiences of the Shrine of Remembrance in Melbourne. As a boy he was both awed and confused by the shrine with its contradictions and complexities such as the melding of the Greek, Christian and pagan ideologies (Inglis 1998). It was the appealing, recognisable and nonthreatening little statue of *The man with the donkey* which brought him a sense of familiarity and belonging. Donkeys are indeed familiar and reassuring as they have lived alongside humans since the time of their domestication, serving as both symbol and utility.[1] The enduring influence of the donkey plays a significant role in memorials to 'the man with the donkey' as an actor in the actual event and as a symbol of humility and service. I argue that the agency and experience of the actual donkey, over and above their use as symbol, is finally being recognised in these memorials.

Simpson and the donkey have become a central focus of Anzac Day, the most significant secular celebration observed in Australia today. Celebrated on 25 April, this national day of remembrance commemorates the anniversary of the landing in Gallipoli in 1915.[2] Out of the horrors experienced at Gallipoli stepped the reassuring sight of the man and the donkey, carrying a wounded soldier to safety. Together, they came to represent the noble Anzac, not the brutality of war but the bravery and mateship of the digger. Furthermore, they symbolised hope and redemption. The mythology surrounding the image of man and donkey is deeply rooted in Christian tradition that acquired a uniquely Australian significance at this defining moment in Australia's history when the foundation was laid for a national identity. The image of man and donkey combined Christian mythology and the egalitarian, working-class bush identity that makes Australia unique: the donkey is instrumental in connecting the religious and the secular. This becomes evident from an analysis of the major memorials to 'the man with the donkey' in Australia and overseas.

To understand the significance of the donkey, however, we cannot look at it merely as a symbol since that would overlook the power and agency of the real animal and reduce its potency to a human projection and fail to acknowledge the donkey as a material actor in the event being memorialised. As Erica Fudge has observed:

Animals are present in most Western cultures for practical use and it is in this material relation with the animals that that representation must be grounded – the historian must analyse the uses to which animals were put. If we ignore the domination of humans we ignore the

---

1   Donkeys have been endowed with various symbolic meanings over their long history in association with humans, see Bough (2011).
2   The campaign fought on the Gallipoli peninsula in Turkey during the First World War was part of a joint British and French operation mounted to capture Constantinople. The campaign was the first major battle undertaken by Australian and New Zealand troops known as ANZACs. All Australians who served and died in all wars, conflicts, and peacekeeping operations are now remembered on Anzac Day.

fundamental role that animals have played in the past. A symbolic animal is only a symbol unless it is related to the real. (Fudge 2002)

To understand the role of the symbolic donkey in the growing Anzac legend, we must first look at the role of the real donkeys at Gallipoli.

Gallipoli

How Simpson's donkey came to Gallipoli we cannot know for certain. He may have been one of the donkeys brought from the island of Imbros as General Birdwood, commander in chief of the Anzacs, claimed (Bean 1921, 573). Concerned about the problem of transporting water to the troops on the front lines, Birdwood obtained about 100 small donkeys from Imbros to carry water in kerosene cans. CEW Bean, the Australian war correspondent, noted that:

> A number of donkeys with Greek drivers had been landed on April 25th for water carrying. The drivers were soon deported, and after the first days the donkeys fed idly in the gullies, till they gradually disappeared. (Bean 1921, 553)

Presumably some of the donkeys were killed by stray bullets but those that remained soon attracted the attention of the soldiers, their freedom and antics a source of diversion between battles.

Donkeys are known to be hardy and steady animals and their surefootedness made them suited to bearing burdens in the rocky terrain and steep paths at Gallipoli. They carted water, food, medical supplies and ammunition. Their placid nature also meant that donkeys offered familiarity and comfort in an environment that was strange and inhospitable to the Anzac soldiers. They also provided companionship, becoming a focus for fun and affection. Some soldiers may have believed that donkeys led charmed lives. As Trooper Bluegum reported in the *Sydney Morning Herald*, 'The soldiers swore by the donkeys [sic] luck, and when the shells burst stood by the animals rather than fly for shelter' (September 1915). Two donkeys, Jenny and her foal Little Jenny, were particular favourites and the men were greatly saddened when the foal was hit and died a slow and painful death. With death and suffering all around them, they felt compassion for an innocent donkey foal caught up in the battle and grieved her loss. She was a victim caught in the crossfire, just as they were. As one soldier recorded:

> Our sorrow is immeasurable. The mother never left her babe whilst it suffered excruciating agony through a deadly shrapnel pellet . . . Jenny Senior is grief stricken and now lies on the neat little grave in which her infant was placed by the big Australian playmates who now mourn their irreparable loss. (Carlyon 2010, 53)

Little Jenny's death and her mother's reaction provided an outlet for the expression of the incomprehensible carnage the soldiers were witnessing all around them. Little Jenny was real as well as being a symbol of the 'innocent' going to their deaths. She was also a victim of human domination: brought to the peninsula for human purposes, she was needed by them yet sacrificed by them.

Donkeys worked with stretcher bearers from the very start of the Gallipoli campaign and it is this role for which they are best known. Michael Tyquin in his history of the Australian army medical services at Gallipoli noted that Simpson began working with donkeys on 26 April, the day after the landing, and that he: 'was one of a number of medics using donkeys for conveying lightly wounded soldiers between gullies and the dressing stations' (Tyquin 1995, 27). The donkeys, recognised as being calm in a crisis, were trained to carry a man up to the operating table and then back out of the tent. New Zealand field stretcher bearers as well as the Australians 'made use of those hardy little animals for those who were lightly wounded' (Carbery 1924, 105).

One of these donkeys was destined for immortality but, as is the case with many legends, he (or they) had several names, most commonly: Murphy (as I shall call him), Duffy 1 and 2, Abdul and Queen Elizabeth. Various accounts of Simpson have him using differing numbers of donkeys; some have him using two at a time and others using different donkeys after one was shot. Simpson himself was known by different names including Simmo and Murphy, all of which added to the mystery and mythology surrounding Murphy and Simpson. John Simpson Kirkpatrick came from South Shields in the UK. Having deserted from the merchant navy, twenty-two-year-old Simpson travelled around Australia working at a variety of jobs. He was a trade union activist and did not take well to authority. However, he enlisted in the Australian Imperial Force (AIF) hoping for a chance to return to England. Instead he found himself at Gallipoli serving in the 3rd Field Ambulance Regiment where he was killed less than four weeks later. In the numerous accounts written about Simpson both at the time and subsequently, much is embellished, imagined or forgotten, depending on the purposes for which it is intended. Nevertheless, those troops who witnessed man and donkey at work trudging along bearing wounded soldiers to safety in the vicinity of Shrapnel Valley had their spirits raised.

Simpson's work with donkeys as a boy in his home town of South Shields was most probably the reason that he recognised their possibilities when he encountered stray donkeys at Gallipoli. As a boy he had helped care for the donkeys and led rides along the beach during the summer holidays; he was familiar with their temperaments and capabilities. Evidence of this experience can be seen in a photograph of him working Murphy at Gallipoli which shows a simple halter and leading rope fashioned from the bandages he carried as an ambulance bearer. For some, it is a matter of contention that Simpson 'abandoned' his unit and worked alone, however, as far as his companion is concerned, this was a sensible move as Simpson bivouacked with the Indian muleteers to ensure that he had food for Murphy. There were slim pickings on the arid ground for those that grazed freely – and Murphy would need stamina for the work he did each day. Carrying a full-grown man on the rocky and uneven paths was no easy task for a small donkey.

The legend grows

From the start of the campaign at Gallipoli, soldiers' letters home recounted incidents featuring the donkeys – one of the few reassuring subjects they could write about. Letters and diaries, reports and newspaper accounts had already started to eulogise 'the man with the donkey' and the legend which was to grab hold of a nation's imagination was born, the little donkey a firmly embedded feature.

The donkey was a little mouse-coloured animal, no taller than a Newfoundland dog . . . When the enfilading fire down the valley was at its worst, and the orders were posted that the ambulance men must not go out, the Man and the Donkey continued placidly at their work . . . At times they held trenches of hundreds of men spellbound . . . Patiently the little donkey waited under cover, while the man crawled through the thick scrub . . . Then the limp form [of the rescued soldier] was balanced across the back of the patient animal, and . . . the man started off for the beach, the donkey trotting unruffled by his side. (*Sun*, 11 October 1915)

We might now question the veracity of aspects of this report, however, from accounts such as this the legend of Simpson and his donkey was already forming on the battlefields of Gallipoli. Back in Australia, the public embraced the image of the man and his donkey. This was a story that could be told; it illustrated the 'acceptable' face of war: mateship, courage and compassion. The reality of the brutality and violence was too horrific to report or to comprehend. The first photograph of man and donkey appeared in *The Sydney Mail* in September 1915 entitled 'Murphy'. The short feature began with the words: 'Who has not heard of Murphy? . . . a very small donkey who was smuggled ashore and was taken charge of by a brave Australian named Simpson'.[3] The donkey is already centre stage in the public imagination.

Murphy's story was soon picked up and embellished. Many soldiers wanted to be part of it and contributed their own anecdotes about the donkey. Such was the interest in the stories of the donkey that many claims were made about Murphy's origins, and his eventual fate. For example, when Simpson was killed by sniper fire, accounts differ as to what happened to Murphy. Some had him carrying on down to the beach with his passenger aboard; others maintained that, after being frightened off by the noise, he returned to stand by his master's side. As to his eventual fate, there are several contradictory accounts,[4] however we will leave him on the beach at Gallipoli, as reported five months

---

3 The photograph later proved not to be of Simpson but of New Zealander Richard Henderson.
4 He was reported variously having been shipped to Egypt, France, Mudros and India with different units following the evacuation from Gallipoli in December 1915 (Cochrane 1992, 159–62).

after Simpson's death in the *Sydney Morning Herald*: 'Murphy is still plodding along, the idol of the soldiers' (28 September 1915).

Concern for his fate is interesting. There was obviously a hope that he still lived and, had he, Murphy would have been the prize exhibit in any propaganda campaign. 'The man with the donkey' was the central focus of enlistment drives back home in Australia, their story a stirring and patriotic one of courage and loyalty. While the donkeys had particular material relationships at Gallipoli, they were also caught in a much wider network of war-related power and politics.

A poem written at the time by an Anzac soldier shows how the qualities of the diminutive donkey of the growing legend had begun to take on symbolic significance. His hardy steadfast valour and noble heart joined with his master's to become the light to lift the spirits of the wounded soldiers. From 'Simpson's Donkey' ('Crosscut' Wilson, *West Gippsland Gazette*, 16 May 1916):

> Was never burden too great to bear,
> Was never too far away
> No road too rough for the tiny hoofs
> And never too long the day . . .
> And the richest prize in the world to him
> was a wisp of a ration of hay.
> Thro' stony creek and unyielding scrub,
> in thirst and in heat and in cold,
> He picked his way with unerring feet
> and a spirit serene and bold,
> And ever the wounded men gave thanks
> for the two great hearts of gold!
> And I think that angels walked beside
> as they marched in their lowly state;
> Day in day out where the red death smote,
> they carried their precious freight –

> And the sunshine's glow is in hearts today
> That else had been desolate.

The 'real' Murphy became less relevant in the expanding mythology: reports about his (and Simpson's) exploits are contradictory and were surpassed by his symbolic role. What he symbolised became of greater importance in Australia's commemoration of the legend, primarily through the link to Christian iconography.

The religious symbolism of donkeys is vital to understanding the ongoing significance of Simpson and the donkey in Australia, largely because of their associations with Jesus in the New Testament. In particular donkeys were associated with the Messiah who, it was foretold, would ride on a lowly ass to indicate the peaceful nature of his ministry.[5] The image of Jesus riding a donkey on Palm Sunday, one of the most enduring Christian symbols in Western culture with its emphasis on bravery, suffering, hope and redemption, is essential in appreciating the significance of the memorials to Simpson and the donkey. Palm Sunday, although a triumphant moment when Jesus is hailed as a king, was also the first day of the last week of his life. The crowds soon turn against him and he suffered and died at the hands of his enemies just as many soldiers in war do. However, for Christians, this is not a message of death and despair but one of hope and redemption: his death was not in vain; he died to save others. This has become a reassuring message for those suffering and dying – and for those grieving at home. As the wreath sent by Annie Kirkpatrick to the memorial of her brother stated: 'He died so that others might live' (*The Argus*, 1937). While no one would claim that Simpson was another Jesus, his association with a humble donkey makes the inference clear.[6] Without the donkey, Simpson would have been no more, or less, special than others who died at Gallipoli.

---

5   Zechariah (9:9) foretells that the king of Zion will arrive not on a symbol of war, a military chariot, but riding a symbol of peace, a humble donkey.
6   Although Benson suggests this in his sycophantic account of Simpson's life (Benson 1965).

Remembering

With the first anniversary of the Anzac landings approaching in 1916, politicians and recruiters promoted the war effort by using Simpson and the donkey as a powerful propaganda tool for enlistment in Australia. A bitter divide grew between loyalists and anti-conscription groups as the brutal realities of war were becoming more widely recognised. Into this divide stepped Murphy and Simpson, a reassuring image amidst the suffering and grief of the First World War. When interest in Anzac Day increased in the 1920s and 30s, Simpson and his donkey were considered appropriate symbols of remembrance and statues were commissioned to commemorate the bravery and mateship of the ordinary soldier. As the Anzac legend took flight, Christian iconography became embedded in Australian memorials to the man and donkey.

Shrine of Remembrance, Melbourne (1936)

A small bronze effigy of Simpson and the donkey stands in the gardens at Melbourne's Shrine of Remembrance as part of the Gallipoli Memorial (Figure 6.1). Sculpted by Wallace Anderson, this was the first of the sculptures of Simpson and the donkey to be erected in the sacred place of a national shrine.[7] The heroic status of Simpson is obvious in this first Australian memorial. The Victorian Governor Lord Hungerford who presided over the unveiling ceremony paid tribute to 'a great Australian hero' and hoped the memorial would be an inspiration to others, 'an example of courage, self-sacrifice and patriotism and a memorial of one who had lived and died for his country' (*The Argus*, 22 June 1936).

From this early memorial, the significance of the image to Australian identity is evident. The mythology surrounding the image of man and donkey is deeply rooted in the British Christian heritage and hence connects the legend with Australia's colonial heritage. Melded in the image is the powerful Christian message of duty, sacrifice and

---

7   In 1935, Wallace Anderson's design for the memorial was selected though a competition organised by the Australian Red Cross (*The Argus*, 6 July 1935); however, it was not until 1968 that it was moved to its present position.

Figure 6.1 Wallace Arnold (1936). *The man with the donkey*. Shrine of Remembrance, Melbourne. Photograph by Phoebe Michaels.

redemption along with the 'digger', prepared for war by his uniquely Australian bushman identity. A letter that appeared in the press as the shrine neared completion in 1933 suggested that Simpson was 'the embodiment of true sacrifice' and called for a memorial by the shrine. It argued that the man with the donkey carrying a wounded soldier on his back would be of great appeal to the rank and file of the AIF (*The Argus*, 18 October 1933). The letter's theme of equality – and familiarity – was picked up by others:

The proposed monument to the immortal Simpson and his donkey would provide a subject of undying interest. One would turn from the stately grandeur of the Shrine to the homely figure of the plain man and his little grey donkey feeling here was a concrete, vivid illustration of the spirit of heroism that underlies the whole conception of our great Shrine. (*The Argus*, 25 October 1933)

The small bronze statue is simply known as *The man with the donkey* as no individual is recognised at the shrine; however, it was the first statue of an individual member of the AIF (Scates 2009). It symbolises the mateship of soldiers as the group commemorates 'The valour and compassion of the Australian soldier'. Simpson stands tall and strong while the little donkey bearing the slumped wounded soldier is tired but determined, his head hanging low under his heavy load as he leads the group forwards.[8] Displaying the Red Cross on his headband, he personifies innocence and humility in the face of man's inhumanity towards man. It is the donkey especially that helps to evoke gentleness and compassion. Here is the embodiment of the Christian heritage married to the ideals of mateship; no matter whether Simpson was Christian or not, his humble yet noble companion made sure that the association was made.

The fact that Simpson used a donkey at Gallipoli drew attention to them both, just as Jesus drew attention when he entered Jerusalem on a donkey on Palm Sunday. Simpson was an ordinary soldier who, on his own initiative, found a way to help his mates. Throughout his ministry, Jesus lived amongst, taught and helped the ordinary working people. Further to these parallels between Jesus and Simpson, the story of the Good Samaritan (Luke 10:25–37) was equally important to the symbolism of the Simpson and donkey memorials resonating as a call for help for those in need. The modest donkey was important in this imagery. Here was the ideal companion for the Simpson of the growing legend, that ordinary bloke who could be called a Good Samaritan and who became a hero in the service of others. This is the Australian version of mates looking out for each other – and the reason why there were calls for a memorial to 'the man with the donkey'.

---

8   Wallace Anderson went to Melbourne Zoo to make sketches of the donkey there in various moods and positions (Anderson 2010).

The path to getting the memorial erected was not, however, always straightforward.[9] Raising sufficient funds was one problem which can be viewed as an expression of conflict in the ongoing development of the Australian identity following the horrors of Gallipoli. One of the main reasons that the funds for a large memorial were not forthcoming was because the sculpture was erected mainly on the initiative of women, the Red Cross and their relatively humble campaign for subscriptions of a one shilling 'mother's tribute' in order to collect funds.[10] This memorial was not to glorify war but specifically to commemorate all the bereaved mothers and wives who were grieving for their loved ones.[11] The emotive and patriotic image of Simpson and his donkey helping the wounded brought comfort to many mothers, wives, sisters and aunts. They were identified with the 'feminine' qualities of caring for the sick, of nurturing rather than destroying. While Simpson typified the pioneer/bushman/digger, seen as the true 'Aussie bloke' with his larrikin behaviour and disregard for authority, the donkey embodied a female voice. The image of man and donkey together personified the evolving nature of the Australian identity.

A revival in nationalism and interest in Australian history following the social, political and economic upheavals of the 1960s saw the legend re-emerge – but with the downplaying of authority and the valorising of the ordinary soldier – and again Simpson and his donkey fitted the bill. With the approach of the 50th anniversary of the Gallipoli landings in 1965, interest in the Anzacs blossomed and 'the man with the donkey' appeared far more widely in the media. Their iconic place in Australian history was assured by the bicentennial year when on Anzac Day 1988, the grand memorial was unveiled in the nation's capital (Figure 6.2). As historian Peter Cochrane asserted, Simpson and the

9   For a full account, see Cochrane (1990).
10   Fundraising by the social elite of Melbourne for a memorial to General Monash was more in line with the masculine ethos of Anzac and therefore far more successful.
11   Miss Philadelphia Robertson of the Red Cross also believed that the sculpture would appeal to children, with their love of animals. The sculpture was to be in the shelter of a 'Lone Pine' planted in 1933, said to have been grown by the mother (or aunt) of a soldier at Anzac who had sent home a cone from the legendary tree (*The Argus*, 24 October 1933).

donkey were now 'prominently and permanently at the nation's ritual centre' (Cochrane 1992, 238).

Australian War Memorial, Canberra (1988)

The impressive, larger than life-size bronze statue of Simpson and the donkey carrying a wounded soldier was finally erected on Anzac Day 1988 and placed to one side of the memorial's front entrance.[12] It offers a scene of compassion rather than death; sculptor Peter Corlett likened it to Christ entering Jerusalem. Corlett (1986) explores the relationships between ordinary people and animals working together in this memorial. The more realistic portrayal of Simpson emphasises the drama of the events as men and donkey make their way to get aid. The soldier's pain and stress is contrasted with Simpson's weary composure. The little donkey is sturdy and calm although he seems to stagger under his heavy load as he leans forward. With the Red Cross insignia on his headband he represents steadfastness in the face of danger on this perilous journey.[13] It is a scene of empathy amidst the violence of war. The more organic and romantic portrayal of the relationship suggests a return to a harmonious prelapsarian relationship with nature and animals after the brutal destruction of world war.

South Shields (1988)

In the same year that the statue was unveiled at the Australian War Memorial in Canberra, one was erected in Simpson's home town of South Shields in the UK.[14] The contrasts with the Corlett statue in Canberra are striking. The significance of national identity and Christian sym-

---

12   Commissioned on the initiative of the former Labor premier of South Australia and veteran of Korea, Des Corcoran, it was designed and built by sculptor Peter Corlett (eg *Sun*, 10 June 1989).
13   Colonel Sutton apparently took off his own Red Cross armband and tied it around Murphy's head. Sometime after Simpson was killed, Major Butler took the armband and kept it with him throughout the war. Diary of Major HN Butler. The armband is held in the AWM REL/21665/, file no. 89/1010.
14   A limited edition 10 cm high reproduction of the sculpture was cast in bronze by Robert Olley, with one being sent to Australian Prime Minister Bob Hawke on the 75th anniversary of Gallipoli.

Figure 6.2 Peter Corlett (1988). *Simpson and his donkey*. Australian War Memorial, Canberra. Photograph by Jill Bough.

bolism so prominent in the Australian memorials are not evident. The setting for the sculpture could hardly be more secular, placed outside a shopping centre and in front of a pub, 'Kirkpatrick's'. Simpson looms large and impressive in this stylised depiction modeled out of brown painted fibreglass. The donkey is relatively small, vulnerable and unhappy looking. There is certainly not the concern for an anatomically 'correct' and realistic depiction of the donkey as seen in the Australian memorials and, as he is not carrying a wounded soldier it is far from clear as to why he is there (Figure 6.3). Although Simpson was English, few in the UK outside Tyneside have heard his story. This statue is about the man, a hero of Tyneside. There is little sign of the religious significance, the symbolism, the mythical hero, or of the iconic status attained in Australia by 'the man with the donkey'.

Two years later, a bronze statue was erected in another overseas city of 'the man with the donkey' with a similar aim, to honour a man, an Anzac soldier, this time from New Zealand. The bronze statue of Henderson and Simpson's donkey, also called *The man with the donkey*, stands outside the war memorial in Wellington.[15] Private Richard Alexander 'Dick' Henderson of the New Zealand Medical Corps supposedly used Murphy after Simpson was killed to continue the dangerous donkey ambulance work.[16] Although the sculpture also depicts 'a wounded soldier being carried off the battlefield on a donkey', the scene appears almost jaunty. The wounded soldier is smiling and chatting to Henderson, who appears untroubled while the donkey is a caricature with his large broad head, overly prominent ears and strange, sleepy eyes. Henderson is not touching him, there is no leading rope: the donkey is not connected to Henderson in any way. While the story of Simpson and his donkey remains embedded in Australia's national psyche, on the other side of the Tasman, few can name his New Zealand equivalent who picked up where Simpson left off and survived the war (Cumming 2008).

Neither realistic representation nor religious symbolism feature in these two overseas memorials. There seems little to connect these sculptures with Jesus and Palm Sunday, or, indeed, the Good Samaritan. It is only in Australia that the religious imagery is prominent and,

---

15 It was unveiled at the National War Memorial on Poppy Day, 20 April 1990. Created by sculptor Paul Walshe, the statue pays tribute to all medical personnel, stretcher bearers and ambulance drivers who served alongside New Zealand troops in wartime.
16 Henderson survived the Gallipoli Campaign and went on to serve in France, returning to NZ in 1918. Sponsored by Oceanic Life for the Royal New Zealand Returned and Services' Association the bronze cast of Henderson and his donkey was unveiled to commemorate the 75th anniversary of the landings at Gallipoli (Cumming 2008).

Figure 6.3 Robert Olley (1988). *The man with the donkey*. South Shields. Photograph by Tracey Ainsley.

significantly, where the portrayal of the wounded soldiers is unrealistic: only those lightly wounded in the foot could have been carried in this way. The depiction of a meaningful and emotional relationship between human and animal and the religious symbolism of the cross on the headband have become an important aspect of the mythmaking surrounding nationhood which grew out of the battlefields of Gallipoli.

Similar to many Western nations, Australia has become largely a secular rather than a religious society, with its own secular rituals and ceremonies. A prime example of this can be seen in the way Anzac Day has supplanted Easter Day celebrations in terms of its popularity with

6 The donkey and Mr Simpson

the general public.[17] The abiding appeal of 'the man with the donkey' is that it represents both; the two are melded poignantly and powerfully in this familiar and comforting image. In the most recent sculpture, unveiled in Adelaide in February 2012, the emotional connection is present but the emphasis is on equality between human and animal as they face adversity together. This memorial reflects a more contemporary understanding of animal and human relationships: the balance is there physically and metaphorically.

Created by South Australian artist Robert Hannaford, the publicly funded memorial is to serve as a symbol to honour the humanitarian efforts of South Australian Defence Force Health Services, representing bravery but not violence. This dramatic memorial is realistic and forceful, less sentimental than the previous Australian depictions. It describes a dangerous journey for those involved: the pain and fatigue of the wounded soldier, the determination of Simpson. The donkey is exhausted; his head touching the ground, he perseveres. Hannaford's interest was in depicting the stoic determination of man and beast together under the stresses of war. He wanted to show the important role of the donkey, as he chose his footsteps carefully while struggling down Shrapnel Gulley.[18] This donkey is not simply bearing a load; the sculpture depicts the tenacity of this beast of burden: tired and tortured, he struggles on, the burden of history on his back.

Conclusion

As humans acknowledge the contributions of animals to their world, including during times of war, it is fitting that Australia honoured Murphy of Gallipoli. The first animal to be honoured in this way in

---

17   On Easter Sunday, 1990, 18 000 Christians marched through Sydney in a procession led by a donkey. In reporting this event in the *Sydney Morning Herald*, Alan Gill said organisers of the 'Aussie Awakening' would like the donkey to replace the rabbit as the symbol of Easter (16 April 1990).

18   Personal communication with Robert Hannaford, 2 November 2011. He loves animals and is careful to represent them anatomically correctly. He used a donkey in his home town as a model.

Australia,[19] Murphy received the RSPCA's Purple Cross at the Australian War Memorial at a ceremony on 19 May 1997. The plaque reads:

> For Murphy and for all the donkeys used by John Simpson Kirkpatrick, for the exceptional work they performed on behalf of humans while under continual fire at Gallipoli during WWI (1915).[20]

The continuing, indeed growing, interest in the donkey and Simpson and their importance as an Australian icon is clear as memorials to them continue to be commissioned. The familiar and appealing image that brought reassurance and hope to millions has continued over the decades. The donkey, with all his symbolic associations – along with his faithful service and hardy characteristics – plays a significant role despite society's changing attitudes towards war, religion and animals. His enduring influence is evident in the memorials of 'the man with the donkey' both as an actor in the actual event and as a symbol of humility and service. The donkey's representation in the Australian memorials increasingly emphasises the human–animal relationship to the war experience and undermines the idea that the only agents in war were human.

Works cited

Anderson R (2010). *Real life portrait: the life of Wallace Anderson*. Newport, NSW: Big Sky Publishing.

Bean CEW (1921). *The story of ANZAC*. Vol. 1. Sydney: Australian War Memorial.

---

19   The second is Sarbi, an explosive-detection black Labrador deployed in Afghanistan by Australian defence forces, awarded in 2011 for 'the courage she has shown serving her country'. However, neither Sarbi nor Murphy had any choice about being involved in human warfare; their 'noble' and 'brave' efforts were entirely a human construct to cover the less palatable truth that they, along with all those other animals involved in war, suffered and died at human hands. For more information visit www.sandralee.com.au/2011/12/ explosives-detection-dog-sarbi-retires-from-the-australian-army/.

20   Meanwhile, Private John Simpson Kirkpatrick, after years of appeal to be considered for the country's highest military honour, the Victoria Cross, has finally been turned down.

Benson I (1965). *The man with the donkey: John Simpson Kirkpatrick the good Samaritan of Gallipoli*. London: Hodder & Stoughton.

Bough J (2011). *Donkey*. London: Reaktion Books.

Carbery AD (1924). *The New Zealand medical service in the Great War, 1914–1918*. Auckland: Whitcombe & Tombs.

Carlyon L (2010). *The Anzac Book*. 3rd edn. Sydney: New South Books.

Cochrane P (1992). *Simpson and the donkey: the making of a legend*. Melbourne: Melbourne University Press.

Cochrane P (1990). Legendary proportions: the Simpson memorial. *Australian Historical Studies*, 24(94): 3–21.

Corlett P (1986). Simpson and his donkey: a proposal. File 89/1234 AWM.

Cumming G (2008). A picture of bravery. *New Zealand Herald*, 19 April.

Fudge E (2002). A left-handed blow: writing the history of animals. In N Rothfels (ed.), *Representing animals* (pp3–18). Bloomington: Indiana University Press.

Inglis KS (1998). *Sacred places: war memorials in the Australian landscape*. Melbourne: Melbourne University Press.

Scates B (2009). *A place to remember: a history of the Shrine of Remembrance*. Melbourne: Cambridge University Press.

Tyquin MB (1995). *Gallipoli: the medical war. The Australian Army Medical Services in the Dardanelles campaign of 1915*. Sydney: UNSW Press.

Williams H (2011). Ashes to asses: an archaeological perspective of death and donkeys. *Journal of Material Culture*, 16(3): 219–39.

7
Howling, haunting and the symbolic dingo

Amanda Stuart

> A sly, treacherous, especially truculent brute is the
> dingo . . . its long-drawn melancholy howls express
> an infinity of canine wretchedness.
>
> *The Australasian Sketcher*, 31 October 1874

> There is nothing so weird as the howl of the dingo in
> the mallee on moonlight [sic] nights especially. They
> appear to be answering one another from every
> point of the compass. Sometimes a whole family join
> together in one horrifying chorus.
>
> *The Australasian Sketcher*, 21 February 1889

The Australian dingo (*Canis lupus dingo*) is believed to have evolved
from the pale-footed or Indian Wolf (*Canis lupus pallipes*) some
6000–10 000 years ago and arrived from East Asia via seafarers some
3800–4800 years before European settlement (Purcell 2010, 12). For a
brief period following European settlement, the dingo appeared in the
colonial visual record as a scientific curiosity. Its likeness was recorded by a range of amateur and professional artists such as John Gould
(Gould 1863, 138–43), however, it was to vanish from the high art[1] re-

---

1   By this I refer to portraiture, landscape and historical events as depicted in
Australian oil paintings of the mid- to late-19th century.

cord during the mid-1800s. Animal portraiture was, however, alive and well during this time as evidenced by the plentiful family photographs which included loyal dog companions or depictions of valiant hunting dogs. The reason the dingo disappeared from the high art visual record remains unclear, but suggests that it was not a proper subject for serious artists.

The dingo resurfaced in the second half of the 19th century, rampaging across the pages of colonial illustrated newspapers. These images were accompanied by strong editorial opinion that implicated the dingo as a troublesome pest and commonly depicted the dingo in the act of marauding livestock. A number of visual strategies and symbolic devices were employed in these dingo depictions, some of which occur with a frequency that warrants their analysis.

By analysing key examples from colonial pictorial newspapers, this chapter identifies some of the principal strategies and devices used by artists of the late colonial era in their depictions of dingoes. The associations and stereotyping inferred by the images are considered, as is the practice of aligning colonial Australian dingo representations with melancholy and derogatory discourses. The audience towards whom such imagery was targeted is identified and the motivations underlying the use of this imagery considered.

The significance of the dingo's howl within the white settler psyche and imagination is contextualised and draws upon on recent analyses of popular colonial texts. The symbolism of the howling dingo is a consistent motif in visual representations from the era. This howl reverberates in the colonial subconscious and along with ravaged flocks and mutilated carcasses defines the only real contact settlers had with the dingo. Actual sightings and encounters were rare in the wild so it is through early artistic renderings and imaginings that dingoes became visible to the populace. These motifs will be teased out and used as a lens to magnify the subtexts of the imagery.

Conspicuous by absence

Dingo representation is not evident in Australian colonial high art imagery and with good reason. Gone was the European painting tradition of still life that became the bread and butter of early convict artists such as WB Gould (1803–53) that glorified the potential of the colonial bounty. The subject matter of mid-19th-century colonial oil painting tended towards portraiture and landscape and in particular, depictions of notable individuals and desirable estates.[2]

Despite a brief dalliance as a fashionable pet in Sydney in the 1830s (Breckwoldt 1988, 90), the dingo was considered vermin by the pastoralists. This contrasts powerfully with the status of its domesticated cousin, the valued European hunting, working and companion dog. The dingo was regarded as an outright enemy and its demise was directly related to the rise of the colonial sheep industry (Breckwoldt 1988, 83–86). From the 44 sheep that arrived with the First Fleet in 1788, sheep populations rose dramatically to 500 000 by 1861 (Bromby 1986, 51). During the second half of the 19th century, wool was on the cusp of outcompeting gold in terms of financial return in Australia and was therefore embraced as the 'golden fleece', a symbol of colonial industrial bounty.

Shooting, trapping, poisoning and fencing were all employed in the fight against the dingo. A bounty system aimed specifically at dingoes was introduced in New South Wales in 1852 and enshrined in law as *An Act to Facilitate and Encourage the Destruction of Native Dogs* (Breckwoldt 1988, 269). Unsurprisingly, dingo hunting became a sport and regional and urban parties were common in the late 1800s.[3] According to CH Eden (cited in Barrett 1947, 105), 'In the neighborhood of Brisbane and other large towns where they have packs, they run the dingoes as you do foxes at home'.

With hostilities at a new height it followed that, despite some short-lived scrutiny in nature illustration, the dingo became conspicuously absent in the fine art genre. From the mid-19th century until Federation, however, it resurfaced, as a fringe dweller between high and low art, in the illustrated newspapers that flourished predominantly in

---

2 For example, see early works from colonial painters such as Joseph Lycett (1775–1825), Robert Dowling (1827–86) and Eugene von Guerard (1811–1901).
3 Shooting, trapping, snaring, clubbing and poisoning of dingoes and their pups were common practices during this time. The new sport of dingo hunting in the colony rapidly emerged as a functional and popular recreation. With firm roots in European traditions, the Australian colonial version was extended to all, and thereby dissolved class boundaries.

Victoria. Media historian and journalist Peter Dowling (1997, 85) considers dingoes to 'epitomise the explosion of imagery brought about by the commercial application of Thomas Bewick's wood engraving process to mass produced, mass circulation, serially issued magazines'.

Photomechanical technology made its debut in Australia in *The Sydney Illustrated* in August 1888 (Dowling 1997, 95). Until this time, as in England, Australian pictorial newspapers employed hand-cut wood engravings that were allegedly reproduced from photographic images. Spurred on by an increased population due to the gold rushes, there was a huge demand for wood and steel engraving artists working to translate the photographic reproductions. According to art historian Bernard Smith (1945, 71):

> The majority of important artists who began their work during the eighties and nineties of last century [19th] graduated through the illustrated periodical. [In addition they] made the country in the second half of the 19th century one of the most important centres of black and white art in the world.

One such image is a print entitled *The haunt of the dingo* by Hugh George that appeared in *The Australasian Sketcher*, a monthly pictorial published by *The Argus* newspaper, Melbourne (Figure 7.1).

Symbolic hauntings

Hugh George's image (Figure 7.1) depicts a chaotic orgy of predation and carnage. The drama unfolds under the cover of night as the unwitting settlers sleep. The carcass is barely visible, obscured by frenzied feeding. Dingoes charge across the shimmering middle-ground grasslands, summoned by the howl of opportunity elicited from the throat of a victorious protagonist, perched on a precipice in the foreground. This grim tableau is enacted under a shrouded full moon.

Three dingoes are so ensconced in scavenging from the carcass that their forms intermingle with the ribcage and chest cavity of their prey. Their collective presence forms a dark arc of visceral, deathly wings that emanate from the remains of their victim. Four more dingoes encircle them, opportunistically picking at scattered bones. The

Figure 7.1 Hugh George for Wilson & Mackinnon (1874). *The haunt of the dingo*, *The Australasian Sketcher*, 31 October.

single howling dingo simultaneously summons and serenades the gorging pack.[4] The taut form of the howling dingo sits neatly on its haunches, with its nose strained skywards. The moon illuminates this alpha creature in full command of its terrain.

George's representation recalls the native dog of earlier colonial impressions, but vividly reimagines its tyranny and opportunism at the settler's expense. These newly arrived settlers, desperate to make good their dominion, remain ignorant of the howl's true meaning, and are destined to live as the eternally unsettled. The drama unfolds, as often is the case in the colonial imagination, under a swollen moon veiled by clouds. These two motifs – howling dingo and full moon – were

---

4   Personal communication with National Parks and Wildlife NSW pest control officer Scott Guthrie on dingo pack hierarchy, which confirms the tendency for alpha dogs to take the first turn at food, 2010.

frequently repeated in the illustrated newspapers.[5] These signifiers un-nerved a fledgling colony, and reinforced the howl of the dingo as an aural menace. The disputed carcass may well be a metaphor for country.

George's rendition remains a pivotal and important image laden with fundamental symbolism. These matters are taken up along with the oppositional ideas of wild and civilised in the imagery of this time in the following examples.

In Figure 7.2, David Syme and Co. match the wits of the cultured, enlightened male shepherd against the elemental dingo, whose sneaky surveillance manifests visually in its attempts to secure easy, vulnerable prey. Familiar symbols associated with dingoes are embedded in this storyboard – moon, night and chaos.

But there are other ways that this imagery might be read. For example, there are strong associations of dingoes with frontiers and wilderness, as seen in the woodcut entitled *Dingoes prowling around a sheepfold* by Mason Jackson (Figure 7.3).[6]

This is a rare glimpse of how the dingo was imagined back in London. It is also an unacknowledged reproduction of a watercolour attributed to Dr John Doyle and ST Gill, from their collaborative sketchbook, made during a visit by the Irish surgeon to Australia and published in 1862–63 (Gill & Doyle 1993). Here the dingoes appear as shadowy creatures at the threshold of wilderness. Poised and prospecting, they are imagined calculating the opportune moment to strike at the domestic outstation. The dingo is perpetually envisaged at the precipice, the boundary of unknown realms. Such strongholds present fine opportunities for the surveillance of an easy meal.

Travelling sheep camp at night by Frederick Grosse (Figure 7.4) depicts a lurking dingo, in camouflage. The central compositional device

---

5   For examples, see *Prospecting at an outstation* in Gill & Doyle (1863); Mason Jackson, *Dingoes prowling around a sheep fold*, *Illustrated London News*, 3 October 1863; *Wild dogs trapped*, *The Illustrated Melbourne Post*, 5 July 1865; Frederick Grosse, *Travelling sheep: night camp*, *Australian News for Home Readers*, 27 July 1863; *Wild dog hunting*, *Australasian Sketcher*, 21 February 1889; *Shepherding*, *The Illustrated Australian News*, 21 August 1886.
6   All woodcut images from illustrated colonial newspapers are courtesy of TROVE (www.trove.nla.gov.au) and Picture Australia (www.nla.gov.au/app/eresources).

Figure 7.2 David Syme and Co. (1886). *Shepherding, Illustrated Australian News*, 21 August.
Caption: 'A gentle reminder; Poisoning native dogs – successful bait; A rush at night – native dogs among the sheep; Death from poison plant; A sensational story'.

focuses the viewer's attention on the carnage about to unfold as the full moon rises on an otherwise idyllic pastoral scene.

Another reading of dingoes can be seen through the lens of mortality and the futility of life's struggle. The focal point of Hugh George's work (Figure 7.1) is the business of the eight dingoes feeding on the carcass. The motif of a skull placed in the foreground is reminiscent of the Dutch vanitas paintings of the 16th–17th century. The function of this symbolic skull association with the dingo is multifaceted. Compositionally, it provides an offset focal point in the foreground that encloses

Figure 7.3 Mason Jackson (1863). *Dingoes prowling around a sheepfold, Illustrated London News,* 3 October.

a vista, and then directs the eye towards the centre of the image. It also aligns popular culture with an acknowledged high art European painting tradition. By adopting the skull as a symbolic device, it transfers to the colony the European-derived sense of futility in the face of mortality.

Walter Hart (Figure 7.5) employs many of these symbolic devices. There is a sweep of movement across the image as the settler, pursued by a rabid pack, races out from the shadowy wilds towards the safe light of home. Amid the calamity, a bleached cattle skull witnesses the scene. In these works, it is implied that the settler's efforts of forging an

Figure 7.4 Frederick Grosse (1863). *Travelling sheep camp at night,*
Australian News for Home Readers.

honest living in this hostile country are doomed. This subliminal message is evidence of the 'melancholy trope' in Australian colonial art, as
discussed by art historian Ian McLean (1999, 131). He considers melancholy as 'a persistent trope used in discourses of redemption, during
the epoch of colonialism' (136).

Art historian Bernard Smith (1960, 176) also refers to this inherent melancholy in Australian colonial conventions and concludes
'even amongst the "native-born" . . . the virgin Australian landscape . .
. still inspired melancholy feelings' (Smith cited in McLean 1999, 131).
McLean (131) considers Smith was:

> Careful not to reduce the melancholy of colonial art to either per-
> sonal homesickness on part of the artist, or aesthetic conceit . . .
> instead seeking out its wider sociological and ideological meanings.

McLean calls for more focus on 'melancholy as an aesthetic conceit
and tropic formation in evaluating the ideological load of colonial art'.
Such melancholic associations are evident in dingo depictions found in

colonial newspapers. The dingo is persistently loaded as a disturbing, nuisance animal and linked with feared territories such as frontiers and wilderness.

Hunter or hunted?

An alternative reading of Hugh George's work (Figure 7.1) is that the viewer is positioned as the vengeful human hunter, cocking their rifle and readying their waddy sticks[7] for a surprise attack on the enemy. These payback scenarios were evident in colonial imagery and included depictions of sheep stealing by Aboriginal people, coupled with the revenge meted out by white settlers.[8] Documentation of sheep stealing

Figure 7.5 Walter Hart (1866). *The dingo or native dog*, *The Australian Home Reader*, 27 July.

---

7   The Dharuk Aboriginal people of Port Jackson, Sydney, define a waddy stick as a heavy club or war stick. It is a term that was adopted in the late 19th century by colonial wild dog/dingo hunting parties (Lucas & Le Souëf 1909, 10).
8   An example of this can be seen in *Dr Doyle's sketch book*, 1863, plates 13 and 14, *The Marauder* and *The Revenge*.

and similes between Aboriginal people and dingoes are rife in the colonial written records (Parker 2006). However, it is important to acknowledge that this occurred on pictorial levels as well. The possibility of a human creeping up undetected, on any of the events depicted in these works is extremely remote.[9] Dingoes are superb hunters with extraordinary senses and do not linger in the company of humans. These images, therefore, belong to the colonial imagination and are rich in symbolic cliché in their depiction of the crimes perpetrated against the European settler.

In George's work, the viewer is simultaneously placed in the position of hunter and hunted. The absent settler is perceived as much of a victim as the unfortunate prey, which becomes a symbol for the failure of the shepherd. The sense of drama is elevated by the depiction of chaos. It is undisputed that the dingo cost the colony precious livestock and often maimed or left sheep to die a slow and painful death.[10] By imagining the havoc wrought by marauding native dog packs, this image endorses the government-sanctioned destruction of the dingo.[11] The assault upon the sheep industry is an assault upon the ideals of colonial expansion. The main function of images such as Hugh George's (Figure 7.1) was to reinforce negative dingo stereotypes, and to justify the extreme and often cruel and unusual methods of exterminating them.

Yet amongst this calamity there is also cool, picturesque calm. *The haunt of the dingo* is framed by arcs of native vegetation, demonstrating George's clear affinity for Australian eucalypt forests. The motif of the trees and low vegetation echo lyrically through the peripheral planes of the composition and lead the eye to the remnant woodlands on the horizon. A familiar two-thirds landscape to one-third sky – the formula of

---

9   Personal communication with NSW and Victorian pest control officers Bill Morris, John Coman, Scott Guthrie and Tommy Kimber, 2008 & 2010.

10   This situation persists today. In their lust for pastoral selections the settlers encroached upon the dingoes' natural territory, and provided the lure for such predation. In all fairness, they must accept the consequences of providing such an opportunistic feed.

11   This occurred via the commonplace poisoning and trapping of native dogs, as well as the bounty system implemented on dingo tails in 1830. Further to this was the broadening of the *Dogs Act 1852*, which intended to 'encourage the destruction of native dogs' (after Breckwoldt 1988, 93).

balance and harmony – settles these components firmly in the viewer's mind, providing a soothing counterpoint to the sense of drama and anarchy within the imaginary realm of wild animals.

According to Robert Dixon (cited in McLean 1999, 153):

> In any picturesque landscape, the spectator is placed behind the picture plane, and the eye is led through the darkness of the foreground vignettes, into the illuminated depths of the painting.

By positioning the viewer as a sightseer the picturesque is reinforced, which ran in tandem to the success of colonialism (McLean 1999, 158). Here then, is a subtle application of those 'melancholy embellishments' (153) that permeated the picturesque in the Australian colonial imagination.

The loneliest howl: unsettling the settler psyche

So with these images the formula is established: the edge of earthly wilderness, nightfall, the full moon, melancholy, bones, and some unruly – if not hysterical – pack behaviour. This formula has long ancestral lines throughout Western philosophy. By adding the symbolic associations of howling, colonial newspapers perpetrated stereotyping that casts a malevolent shadow around the dingo's demeanour.

From its earliest contact with Europeans, the dingo was aligned with the northern-hemisphere wolf and in many ways became its antipodean equivalent. Barry Lopez (1978, 10) observes, 'there has never been any evidence that wolves howl at the moon, or howl more frequently during a full moon'. Nevertheless this motif has imprinted itself upon the imagination and manifested into countless wolf mythologies and folkloric tales. The European paradigm is transferred to *The haunt of the dingo* (Figure 7.1), where the animal is represented delivering its 'most hideous and melancholic howling' (Skottowe 1988, 35) under a plump, lunar orb. The persistence of this illuminated howl in popular colonial imagery begs the questions, 'why?' and 'for whom?'

The dingo's howl occupies a potent position in the colonial imagination. It was repeatedly described as a dreadful and melancholy moan, one that evoked fear and loathing in humans (Tench 1961, 269;

McCarthy 1963, 469). This was particularly evident in 1790 west of Sydney, during a time when convict deaths were so common that an open pit for the disposal of bodies was situated at Toongabbie. Personal documents attest that:

> At night, the howling of dingoes could be heard around the pit as they fought over the bodies, gnawing at them until they were covered over. (Breckwoldt 1988, 452)

Such a gloomy scenario would serve to deeply entrench fears of being predated upon within the settler psyche. This would have struck deeply at the vulnerability of humans and their will to survive, and no doubt contributed powerfully to the contempt shown towards the dingo in the 19th century, regardless of the dingo's propensity for mutton.

Similar negative stories persisted throughout the expansion of the frontier. Whilst on an expedition to the interior in 1844, English explorer Captain Charles Sturt (cited in Walters 1995, 29) described the desert dingoes he observed as 'rousing us by as melancholy a howl as jackal ever made; their emaciated bodies standing between us and the moon' and as being 'the most wretched objects of the brute creation'.

This damning account by a figure of British authority involved in the expansion of colonial frontiers would have further reinforced the dingo's lowly status. The editorial comment which accompanies Hugh George's *The haunt of the dingo* reads:

> Few persons who have made acquaintance with the Australian dingo or native dog, will readily forget the impression caused the first time they heard the chorus of long drawn melancholy howls in which the animal seems to express an infinity of canine wretchedness. Especially is this the case if the wayfarer was alone by night, perhaps camping by himself in a dark, weird forest, from the depths of which at intervals issue these lugubrious yells, as of lost wailing. (*The Australasian Sketcher*, 31 October 1874)

The disorientating impact of such a call would not have gone unnoticed by the lone shepherd, known to spend weeks at a time in isolation (Bean 1945, 36). The fear of dingoes attacking sheep had a real and constant presence. A firsthand account by Alexander Harris in 1847 described the miserable and solitary life of a convict or free shepherd as being full of floggings and lost wages from merciless masters for lost or injured sheep, 'whether he was right or wrong' (Harris 1977, 182).

Agricultural historian Robin Bromby (1986, 54) attests that

> For most of the settlers, selection came to mean a life spent in poverty, misery and abject loneliness . . . squalor was the norm . . . men would escape to nearby taverns and drink themselves to oblivion, but for the women, there was no escape. Many went mad.

And in his 1945 classic tale of a road trip, journalist CEW Bean (1945, 37) describes the descendants of these remote country shepherds as 'men almost wild – often old hands – that is to say survivors of the far gone days of convict labor in the early colony'.

The settler came to align the dingo's howl with grief, longing and loneliness. This theme is developed by Parker (2006) who reflects on the settler's solitary life and considers how crucial it was in the misinterpretation of the dingo's howl. It was through such isolation that the 'vast emptiness of the land blended with the emptiness of men's personal lives into one long drawn out howl' (Parker 2006, 69). Her analysis focuses on late 19th- and early 20th-century popular literature and poetry published in *The Bulletin* or 'Bushman's bible' as it was known. It was a repository for popular colonial fiction and home to some of the canons of Australian folkloric literature including Banjo Paterson, Henry Lawson, Miles Franklin and Norman Lindsay (Haigh 2008).

Parker (2006) draws on *The Bulletin*'s dingo fictions, which outlined imagined encounters and events between dingoes, humans and domestic dogs. She illustrates that through these stories, largely the work of white, male authors, the howl is aligned with intense human loneliness and unrequited love. For example, in a story called 'Bare fang' by Ion Idriess that was published in 1928 in *The Bulletin*, the howl is interpreted as 'the bottled hopes of a thousand years drifting to hell' (Parker 2006, 69). Parker concludes these fictions to be evidence of the misery of the white settler, which is projected onto the howl and imprinted onto a popular audience. In essence, she implies the misery of the human condition at the frontiers of colonial expansion is transferred to the dingo via its howl, which becomes a metaphor for human sexual frustration and the search for a mate. In Parker's words,

The friendly communication of the dingo is perceived by colonial men as a wail of fear and loneliness or a howl of sexual frustration, in a dual extension of the male settlers' physical and emotional isolation. Bush life created an unnatural situation, where men outnumbered women by twenty to one. Frustrated men interpreted the dingo's howl as the cries of women, especially lovesick women, calling through the dark to the listening male. (2006, 73)

Through popular literature, Parker (2006) identifies the female dingo as a symbol for sexual promiscuity and predation in the form of an obedient domestic dog. In doing so she confirms the dog and human metaphor, by suggesting the domestic (or cultured) dog to be a symbol for the white male settler. The dingo female is thus implicated as a dangerous seductress, a deceitful trickster[12] that toys with her conquests. And so a further discourse of deceit becomes subliminally encoded within the colonial dingo narratives and discourses.

Are there other interpretations of the dingoes' howls? Breckwoldt (1988, 162–63) considers the dingoes' howl to be 'wilderness in sound', and acknowledges a wide range of human responses:

For some, the howl of the dingo is sinister and eerie. Others find it friendly and melodious. There are romantics who, deprived of lion, tiger or jaguar, cling to the howl of the dingo as something wild and primeval in a bush that is so easily passed off as a monotone of eucalypts and cryptic marsupials.

Bill Neidjie, an Aboriginal man born in Oenpellie mission in approximately 1911, interprets the dingo howl as a sociable form of distance communication, which relays useful information about food, courtship and family business (Parker 2006, 71). Australian scientists Laurie Corbett and Alan Newsome (Corbett & Newsome 1975, 369–79) agree

---

12 Calling up or imitating a dingoes' howl is a traditional way of ensnaring a dingo, used by dog trappers, implicating *them* as the deceitful tricksters. This appears to turn the colonial literary cliché on its head. When penning stories regarding the promiscuity of the female dingoes, perhaps the authors were drawing on a uniquely human male way of luring dingoes in search of a mate to their fate.

with this interpretation and consider that an animal such as the dingo that can disperse and reform into groups when required for cooperative hunting, breeding and raising of young must have a sophisticated communication system.[13] Although they acknowledge the full tonal range of the howl is not yet adequately studied, they identify its functions: locate a mate, locate members of same group for cooperative hunting, define territory, avoid strangers, call pups and signal members of same group they are moving.

Conclusion

This discussion has illuminated the smorgasbord of symbolism found in popular illustrated 19th-century colonial newspapers. When imagery such as Hugh George's appeared in these newspapers, it was entangled in emotive and disparaging opinions about dingoes. The intention was to drive home the clichés, mythologies and negative stereotypes associated with dingoes to a burgeoning middle-class and increasingly urban audience. Once these were embedded, the unquestioned implementation of dingo eradication strategies could be justified. These were essential to the advancement of the colonial agenda, which was specifically, in this case, to preserve the golden fleece. In journalist historian Peter Dowling's words:

> The appeal of illustrated newspapers resided in their capacity to provide a predominantly urban-dwelling, middle-class readership with visual confirmation of its fervent belief in colonial progress. (Dowling 2008)

---

13   The dingo is not a highly vocal animal but possesses three types of vocalisations that are recognised by humans: the moan, the bark howl and the howl. Of these, the howl is thought to have considerable subtlety. Howling enables long-distance communication and serves to locate others in the group, and to identify separate groups thereby avoiding confrontation. The frequency of howling increases in breeding season as it is used to attract or select a mate. It also aids in group cohesion and avoiding strangers. The technique of 'howling up' or imitating a dingo by trappers works best in mating season (Breckwoldt 1988, 163).

Images, combined with strong editorial opinion that championed colonial advancement, were designed to influence and appeal to a middle-class, urban population.

The sheep or cattle carcass is repeatedly the target of the dingo. Domestic livestock symbolises the toil of the honest and hardworking settler. Following the gold rushes, the sheep industry – as the financial backbone of the colony – became a popular subject in illustrated newspapers and a favourite of the Australian impressionist movement. Many artists who were apprenticed to the illustrated pictorials emerged late in the century as fully fledged artists of the Heidelberg movement. This high art painting movement bristled with nationalistic fervour and was fuelled thematically by the celebration of rural pride in the lead up to Australian Federation. It is no surprise that under the helm of this next wave of nationalistic artists the native dog would again vanish from the colonial visual record.

Acknowledgements

The author would like to acknowledge Craig Mills and thank him for his exceptional editing contribution. Without his efforts at the 11th hour, this work would not have been possible.

Works cited

Barrett C (1947). *An Australian animal book*. Melbourne: Oxford University Press.
Bean CEW (1945). *On the wool track*. Sydney: Angus & Robertson.
Breckwoldt R (1988). *A very elegant animal, the dingo*. Sydney: Angus & Robertson.
Bromby R (1986). *The farming of Australia*. Sydney: Doubleday.
Corbett L & Newsome A (1975). Dingo society and its maintenance: a preliminary analysis. In MW Fox (ed.), *The wild canids: their systematics, behavioural ecology and evolution* (pp369–79). New York: Van Nostrand Reinhold.
Dowling P (2008). Illustrated newspapers. eMelbourne: the city past and present. [Online] Available: www.emelbourne.net.au/biogs/EM00741b.htm [Accessed 23 April 2009].

Dowling P (1997). Destined not to survive: the illustrated newspapers of colonial Australia. *Studies in Newspaper and Periodical History, 1995 Annual*, 1997: 85–98.

Editor (1889). Wild dog hunting. *The Australasian Sketcher*, 21 February.

Gill ST & Doyle JT (1993 [1862–63]). *Dr Doyle's Sketchbook*. Revised edn. Sydney: Mitchell Library Press and Centaur Press.

Gould J (1863). *The mammals of Australia*. Vol. 3. London: Taylor and Francis.

Haigh G (2008). Packed it in – the demise of the *Bulletin*. *The Monthly*. [Online] Available: www.themonthly.com.au/issue/2008/march/1268869044/gideon-haigh/packed-it [Accessed 7 August 2013].

Harris A (1977 [1847]). *Settlers and convicts: recollections of sixteen years' labour in the Australian backwoods – by an emigrant mechanic*. Melbourne: Melbourne University Press.

Lopez B (1978). *Of wolves and men*. New York: Scribner.

Lucas AHS & Le Souëf WHD (1909). *The animals of Australia: mammals, reptiles and amphibians*. Melbourne: Whitcombe and Tombs.

McCarthy PH (1963). The dingo – a comment on an early reference by the Spanish navigator, Don Diego De Prado Y Tovar. *Australian Veterinary Journal*, 39: 469–70.

McLean I (1999). Under Saturn: melancholy and the colonial imagination. In N Thomas & D Losche (eds), *Double vision: art histories and colonial histories in the Pacific* (pp131–58). Cambridge: Cambridge University Press.

Parker M (2006). Bringing them home: discursive representations of the dingo by Aboriginal, colonial and contemporary Australians. PhD thesis. University of Tasmania.

Purcell B (2010). *Dingo*. Collingwood: CSIRO publishing.

Rolls E (1969). *They all ran wild: the animals and plants that plague Australia*. Sydney: Angus & Robertson.

Skottowe T (1988 [1813]). *Selected specimens from nature of the birds, animals & C. &C. of New South Wales*. Edited by T Bonyhady. Sydney: David Ell Press, Horden House.

Smith B (1960). *European vision and the South Pacific, 1768–1860: a study in the history of art and ideas*. Oxford: Oxford University Press.

Smith B (1945). *Place, taste and tradition*. Sydney: Ure Smith.

Tench W (1961). *Sydney's first four years: being a reprint of a narrative of the expedition to Botany Bay and a complete account of the settlement at Port Jackson*. Sydney: Angus & Robertson.

Walters B (1995). *The company of dingoes*. Australian Native Dog Conservation Society, Rozelle: Standard Publishing House.

8

Animal approximations: depicting cryptic species

Anne Taylor

The categories of animal and human are no longer separated by a rigid dichotomy. Their disjunction has been called into question by the contemporary expansion of knowledge in the fields of animal cognition, communication and social behaviour. In philosophy, human subjectivity has been recast as a heterogeneous and mobile pluralism, characterised by an acceptance of difference and the recognition of multiplicity within us. Yet animal rights discourse consistently employs the figure of the singular, charismatic animal in the fostering of an ethical responsibility towards animals (Daston & Mitman 2005, 10). We seem to require ethical relationships with animals that mimic those we have with humans. The more familiar we are with particular animals, the more likely we are to empathise with them and extend to them ethical consideration (Calarco 2008, 8–9). It is difficult to feel deep concern for sea creatures that are little more than animate slime, for example, or microscopic organisms that cannot be observed below unassisted levels of detection, even if they are crucial to maintaining environmental stability. Nevertheless, even marginal creatures inhabiting unexplored or inaccessible environments are threatened by human technologies and interventions that intrude upon animal life on a vast scale. A habitual anthropocentrism blinds us to the consequences of our ever-increasing manipulation of the natural world and its nonhuman inhabitants.

In considering the role of visual depiction in promoting understanding of the beauty, complexity and vulnerability of cryptic or invisible life forms, this chapter presents an argument for ethical generosity towards all levels of animal life. It outlines a selected visual history of recondite marine animals and microscopic marine life, choosing examples that demonstrate changing historical attitudes towards unfamiliar species and a range of anthropocentric tendencies. Early depictions were purely anthropocentric, intended to symbolise aspects of human experience. Scientific depictions are relatively recent, making their first consistent appearance as observational records of discovery in the 18th century. Initial scientific drawings of unfamiliar species were later incorporated into fine art and design, taking on meanings that reflected aspects of human relationships with the natural world. In contemporary art, through a critical aesthetics of recombination and recontextualisation, the depiction of cryptic species is given imaginative and hybrid forms, probing the interaction of improbable animals and enhanced humans, often in response to social and environmental concerns. The illustrative, documentary and inventive aspects of visual depictions establish an influential awareness of cryptic animals that increases our understanding and valuing of their existence. As the art historian Bernard Smith contends in his survey of the visual records produced during Cook's voyages of exploration, 'Words are often forgotten but images remain' (1992, 193). Depictions of unfamiliar species in contemporary art destabilise anthropocentrism by employing an increasingly complex interplay of mythology, science, ethics and aesthetics, and initiate an empathetic alignment between human concerns and hidden animal lives.

Ethical generosity

Although all animal lives are largely mysterious to humans, there are realms of the animal world that are invisible to human imagination, existing beyond the horizons of our experience. Our indifference to vast hidden tracts of animal existence impoverishes our awareness of the meshed connections linking living beings. The biological sciences have penetrated even the most resistant environments, yet their findings are fragmented and incomplete. The hidden niches of the world's forest, desert and marine wildernesses nourish animals ingeniously adapted to extremes. Many wilderness or deep-sea species elude comprehensive surveillance, disappearing into the vast, untracked topographies of the wild. Microscopic creatures are only visible through the manipulation of lenses and slides, isolated from their teeming habitats. In a difficult but necessary extension of post-human ethics, we have a responsibility not to diminish these remote existences but instead protect them as we would more familiar creatures.

The philosophic traditions which provide a basis for thinking through an ethics inclusive of animals are grounded in an essential anthropocentrism that makes the moral standing of animals difficult to delineate. The ethical relation proposed by Immanuel Kant, which still exerts a strong influence today, assumes a universal subject difficult to apply to human difference much less to the multiplicity of animal species (Wolfe 2003, xii). Even a pluralistic approach to animal ethics can proceed as if final hierarchic judgments are possible, closing off the infinite permutations of alterity. Instead of attempting to delineate normative demarcations, we need to adopt an ethical openness that allows inclusiveness, in spite of the risk of absurdity (Calarco 2008, 72).

The many species of animals whose strangeness or formlessness make them unattractive or invisible to human sympathies nevertheless perform important roles in the maintenance of a flourishing environment. Subterranean creatures function as agents for fertile and friable soils, burrowing blindly beneath our feet. Coral gardens, a symbiotic interface between animals and plants, provide living shelters for a multitude of sea creatures as well as protective ramparts for shorelines and islands. The cryptic inhabitants of these coral labyrinths, fleshy, spined, lumpen, or gelatinous, add to the trophic diversity of the reef's biomass. Drifts of zooplankton contribute to the basis of nutritional pyramids for all marine life. Anthropomorphic identification fails in these cases, and we feel instead a sense of disconnection, even if their forms are beautiful. These unfamiliar species are often described as alien creatures, likened to inhabitants of far galaxies or the phantasms of horror stories. We find it difficult to extend to them the empathetic alignment with human experience assumed for domestic pets or charismatic mammals. These forms of life are nevertheless extremely vulnerable to human technologies and interventions, which now operate on a global scale, threatening whole ecosystems, including those beyond human awareness.

The difficulty of allowing ethical consideration for such marginal creatures is obvious and can only be justified through a broad recognition of the value of all forms of life (Attfield 2003, 44–45). It is not the individual value of these creatures as singular, independent entities that makes their protection necessary, but their contribution to holistic, environmental interactions that nurture and sustain the life-world. The kinetic self-organisation of existence, generated through the interplay of entropy and cohesion, transformation and stability, is enacted on a multitude of levels, whether visible or hidden. The philosopher Elizabeth Grosz identifies this biological complexity as a 'contrary current' to the entropy of inorganic processes, instead 'producing new dynamism in the universe, a force of self-proliferation' (2004, 200). The evolutionary impetus which creates diversity is located in the mutability and responsiveness of biological development across all forms of life. Often the importance of these processes remains veiled to human perceptions, and it is possible that there are systems as yet undiscovered, which may only come to light if they are disrupted. The diminishing variety and fecundity of a world overburdened with technological manipulation reduces natural resources, but also leads to the impoverishment of human potential for intellectual and creative development.

Symbolic animals

Though the political and economic organisation of the industrialised societies remains largely insensitive to the entangled processes of biology, a groundswell of ecological awareness has begun to infiltrate the mechanistic and instrumental thinking of governments and corporations. The role of striking imagery of animals and their habitats, particularly in the form of documentary photography, has been central to the fostering of empathetic sensitivity towards the natural world. Understandably, these images have been largely focused on appealing creatures and picturesque environments. Although popular scientific imagery and video includes cryptic species, the mediated presentation of these genres reassures the audience with a concocted narrative coherence. Investigations into difficult habitats often emphasise heroic examples of human ingenuity and technological inventiveness rather than focusing on exploring unfamiliar animal lives. Storylines, voiceovers, musical accompaniments or didactic texts direct our responses, giving an illusion of novelty while tailoring the information to popular taste. An aesthetic of identification engages our sympathies and encourages compassion, but a more nuanced approach to eco-imagery should accompany such direct anthropomorphic appeals to the emotions.

These stylised depictions are reinforced by the traditions of Western imagery, particularly in the representation of European animals, which were given deeply symbolic significance. Animals appearing in medieval and Renaissance illuminated manuscripts, bestiaries, and paintings were endowed with anthropomorphised temperaments and characteristics. The early bestiaries were not zoological texts, but divine exegeses and vehicles for the teaching of Christian values (Baxter 1998, 136). The depiction of symbolic animals provided metaphors for human foibles and virtues, enacting vivid somatic portrayals of moral behaviour. Transferring moral tales from human to animal protagonists drew on stereotypes of animal propensities to clarify the moral lessons offered. Less familiar creatures, such as marine dwellers, were viewed with revulsion, given cold or dubious qualities, or shown as hybrid monsters (Kemp 2007, 2). Often these spiritual or demonic mythological beings were part human, and represented the seepage of an otherworld into everyday existence. World maps showed unexplored oceans as the haunt of chimerical beasts that were elaborated from mythology and the fanciful descriptions of sailors and explorers (Figure 8.1).

The affluence of a prosperous European bourgeoisie and its successful exploitation of sea power to colonise exotic lands produced an alternative depiction of marine life as glistening bounty in still-life paintings of rich feasts (Bryson 1990, 105), which focused on the human world rather than the animal. As a more exotic form of still life, seashells collected from the tropics were valued commodities, not for their status as housing living creatures, but for their decorative and convoluted forms that suggested artificiality (Bryson 1990, 109).

SEA-SERPENT ATTACKING A VESSEL. FROM OLAUS MAGNUS.

Figure 8.1 *Sea serpent from Olaus Magnus, 1522* (1895), *Strand Magazine,* July–December.

Creatures of the deep

The scientific paradigm engendered by the rationalism of the Enlightenment rejected the sympathetic alignment of human and animal. Anthropocentric attitudes manifested instead as a hierarchic ordering of species, with 'man' as its pinnacle, imposing a taxonomic grid of description and comparison on the natural world, and focusing on its instrumental and economic uses (Foucault 1994, 137). Attention was directed towards the mapping and exploration of newly discovered lands. The ocean journey was considered a hazardous passage between harbours and once established, sea routes were generally adhered to. Mariners whose livelihood depended on hunting and fishing were more adventurous, particularly whalers and sealers who needed to track their prey beyond familiar routes. The knowledge of the ocean supplied by sperm whalers, who ventured into unexplored tracts of ocean, formed the basis of early oceanography (Rozwadowski 2005, 40). The accurate representation of remote marine life was a neglected field until these voyages of discovery fostered more concentrated investigation into all aspects of the ocean.

The exploratory voyages of James Cook initiated a scientific approach to geographic discovery, including detailed recording of biological, mineral and topographic aspects of the new lands and seas (Smith 1992, 51–52). Artists' skills were essential for recording finds, and specimens were drawn and painted with minute attention to detail. Today, the fragile awkwardness of some of these scientific illustrations is part of their charm, registering the unfamiliarity of the newly discovered flora and fauna. Many marine species were difficult to preserve permanently, so accurate field drawings were an integral avenue of study.

The practice of observing or depicting marine organisms in their natural settings underwater was only conceived after the keeping of aquariums became common in the mid-19th century. The terrestrial view is standard in earlier illustrations and paintings showing marine life stranded awkwardly on the shore, swimming on the surface of the ocean or decoratively arranged in garlands (Gould 1998, 67). The aquarium allowed a side-on view through clear glass to become naturalised, advancing the understanding of sea creatures as graceful, swift and perfectly adapted to their environment. Invertebrates appeared shapeless and limp out of water, but underwater flowered into supple forms in drifting motion (Figure 8.2).

Some of the most beautiful examples of marine illustrations for scientific purposes were produced on the Baudin expedition, which left from Le Havre in 1800 to explore the Antipodes. The artists Charles Alexandre Leseur and Nicholas-Martin Petit recorded and collected biological specimens including fragile sea creatures under the direction of the zoologist Francois Peron. Peron commented in his journal:

> long has the study of mollusks and soft marine animals been neglected by naturalists and explorers. Yet some of these animals, in their exotic shapes, unique structure, beautiful colours and variety of habitat, richly deserve the attention of the enlightened community. (Peron quoted in Hunt & Carter 1999, 11)

Without breathing apparatus for deep-water diving, remote methods of collecting marine species had to be devised. Scientific dredging, a technique adapted from oyster dredging, scoured the ocean bed and raised specimens to the surface. These operations were at first carried out in shallow waters by amateur scientists working from rowboats and private yachts. In the 1870s dredging techniques were adapted to ocean-going ships exploring and sounding the deep ocean, making

Figure 8.2 Frontispiece from Jabez Hogg (1865). *The microscope: its history, construction, and applications*. London: H. Ingram and Co. Underwater view with enlarged microscopic organisms.

possible close viewing of many previously inaccessible animals (Rozwadowski 2005, 168). Microscopes were used to examine minute amounts of bottom sediment or samples of sea water (Figure 8.3).

It was debated whether life could exist at great depths as substantial evidence was difficult to obtain. Eventually definitive proof was provided by the raising of a failed submarine telegraph cable from a

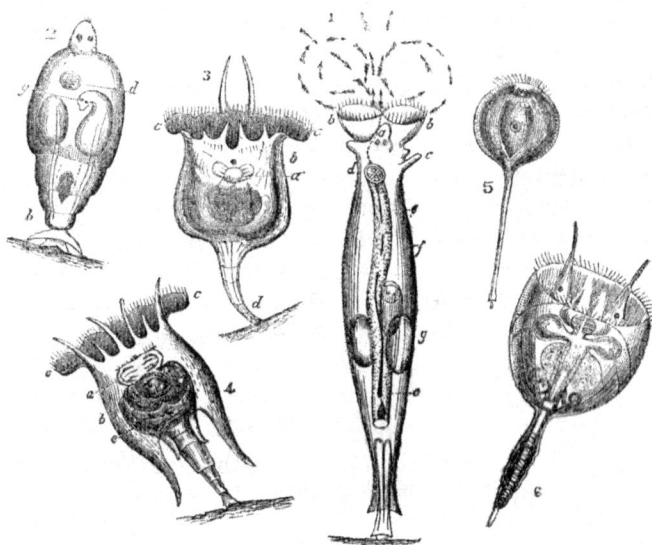

fig. 101.

1. The common Wheel-Animalcule, with its cilia or rotators pointed. 2. The same in a contracted state at rest: at *g* is seen the development of the eyes in the young. 3. Pitcher-shaped Brachionus: *a* the jaws; *b* the shell; *c* the cilia, or rotators; *d* the tail. 4. Baker's Brachionus: *a* the jaws and teeth; *b* the shell; *c* the rotators; *e* the stomach. 5 and 6. Other forms of the same family.

Figure 8.3 Microscopic infusoria from Jabez Hogg (1856). *The microscope: its history, construction, and applications.* London: H. Ingram and Co.

depth of 1200 fathoms, which was found encrusted with marine organisms (Rozwadowski 2005, 141). Interest in deep-sea creatures as a new field of study prompted more focused scientific voyages, most notably that of HMAS *Challenger* in 1872. Specimens, drawings and reports from this voyage were disseminated widely in scientific circles, helping to establish oceanography and marine biology as a serious discipline (Rozwadowski 2005, 168).

The German zoologist and artist Ernst Haeckel was one of the interpreters of the *Challenger* data, editing the material on medusae, siphonophores and radiolarians[1] (Breidbach 2005, 22). As a supporter of Charles Darwin, Haeckel's illustrations were not confined to recording the precise details of unfamiliar creatures but were ordered in their variation of natural forms into a visible translation of the evolutionary process (Breidbach 2004, 9). For Haeckel, the aesthetic forms of biological specimens, even those of microscopic organisms, embodied the hidden laws of the natural world. The symmetries of repeated forms, accumulating into the complexity of advanced life forms, were interpreted as a system of diversity expressed through ornament. Evolution acquired an aesthetic articulation through the crystalline or sinuous patterning of natural structures, displaying a biogenetic dynamism in the visible forms of organisms which moderated the mechanistic, anthropocentric view of nature adopted by pre-evolutionary science.

Visible forms were required for the reconstruction of evolutionary processes. However, dissemination of marine invertebrates for study and museum display proved difficult as their soft bodies deteriorated and preservation in alcohol resulted in loss of colour and shape. Haeckel's marine illustrations contributed to the design of detailed glass models fashioned by Leopold and Rudolph Blaschka. From 1836, they constructed intricate replicas of marine creatures using Bohemian glassblowing techniques. Today the exquisite but fragile models are being restored, valued for their scientific accuracy as well as for their beauty and skilful construction (Albert R Mann Library 2009).

Haeckel's influence also extended to the arts and architecture. He presented his scientific drawings as aesthetic templates in the publication *Art forms in nature* (1904), which were elaborated upon by René Binet who worked them into designs for the decorative arts by adapting the stylised shapes of microscopic and marine animals to ornamental objects. Undoubtedly Haeckel's images, as well as his theories, influenced those designers who produced the whiplash arabesques and tessellated geometries of *Jugendstil*,[2] (Proctor 2007, 10). But Haeckel also adapted his initial drawings of preserved specimens to illustrations that conformed to the ornamental style of his times (Breidbach 2004, 14). In the interaction of direct scientific observation and the adapt-

---

1   Medusae are jellyfish, siphonaphores are translucent marine organisms with simple, tube- or cup-shaped bodies, and radiolarians are microscopic marine organisms with a latticed skeleton.
2   This is the German term for Art Nouveau, an artistic style most commonly used in the decorative arts and illustration, appearing in the 1880s until the start of the First World War.

ation to the fashionable Art Nouveau style, translating jellyfish into sinuous floral ornaments, or the skeletons of radiolarians into latticed symmetries, Haeckel produced images which caught the imagination of a wide audience. He successfully disseminated awareness of these strange and cryptic marine creatures through streamlining their alien forms into accessible, aesthetic images. The serious biological theories on evolutionary form embodied in the plates of *Art forms in nature* were, however, overshadowed by the exotic appeal of newly recorded species and their exaggerated, ornamental symmetries (Figure 8.4).

As modernism evolved away from a fascination with decorative, free-flowing forms into biomorphic and geometric abstraction, figurative imagery was elided into subconscious association and perceptual slippages. References to the animal world became metaphors for psychological or emotional states. For the surrealists, animal figuration suggested Freudian evocations of subconscious states fluctuating between desire and dread (Foster 1993, xvii–xviii). Max Ernst's settings of coralline encrustations, created through the contingent manipulation of materials, evoke both the protean mutations of geological and biological energies and the shifting depths of the subconscious (Foster 1993, 21–25).[3] The surrealist interest in evocative objects, including those from Oceanic cultures, was extended to natural entities that suggested the exotic and the supernatural.

Imaginary ecologies of race and feminism

Contemporary images of animals overlay these psychosomatic metaphors with social, political and ecological concerns, though depictions of cryptic species remain relatively rare. In the series *Watery ecstatic*, begun in 2001, the African-American artist Ellen Gallagher retains subjective associations but complicates her references to the animal world with investigations into gender and racial issues. For Gallagher sea creatures suggest the metamorphosed progeny of African slaves, mostly women and children, who were cast overboard during the dangerous Middle Passage to the New World. She incorporates repeated,

---

3   See Ernst's paintings *The eye of silence*, 1943–44, oil on canvas; and *Europe after the rain II, 1940–42*, oil on canvas.

Figure 8.4 Ernst Haeckel (1904). 'Trachomedusae', plate 26 in *Art Forms in Nature*.

but discreet, images from vintage black magazines so that the marine organisms are subtly embedded with signifiers of black identity. Collaged cutouts of wigs and the elaborate curls of afro-style hairdos morph into the tentacles of marine creatures. Cartoon lips and eyes replace the polyps and suckers of corals and octopuses, enlaced with seaweed arabesques. The images from the 1930s are used as a critique of racial stereotypes and to capture the traces of real lives that are embedded in the past, but retain their specificity (Gallagher 2005b).

Gallagher's mythical sea dwellers are created using traditional watercolours, often by isolating the creatures on a plain background in the same style that naturalists used for scientific illustration. The artist spent time as a student on an oceanographic research vessel, studying and drawing microscopic wing-footed snails. She also develops novel techniques of cutting and carving into the thick paper, sometimes creating a sliced and textured surface that requires a close-up view to reveal details. Gallagher reminds us that in water all dead things turn white, evoking a ghostly trauma. The monochrome paper suggests the erased history of the black slaves, their drowned bodies excised from the archive (Malik 2006, 30–34). The artist compares these methods to scrimshaw carving – the whaler's pastime of scratching designs into bone (Gallagher 2005a). The tedious and difficult voyages of working sailors are recalled, along with the longing and deprivation that prompted their tales of monsters and mermaids. The prosperity of the Western world was built on these lonely explorations and far-flung colonisations, and the wealth generated by slave labour. As a direct inheritance of this exploitation, industrial and technological empires have proliferated, degrading the human world as well as the habitats of wild creatures.

My own paintings, drawings and prints draw on historical and contemporary imagery to explore the interaction of human and natural worlds, engaging with feminist and ecological concerns. I depict marine environments, often difficult or remote, where to survive, humans must adopt prosthetic equipment that echoes the streamlined, tentacled or finned forms of sea creatures. Scenarios of watery environments colonised by intrusive technologies suggest a future that will include environmental uncertainty and a need for humans to respond more sensitively to the natural world (Figure 8.5).

I hope to initiate an engagement with life forms that exist beyond our ordinary level of experience, creating metamorphic resonances with the inner realms of the human body, the senses and the psyche. The convoluted, fleshy forms of marine invertebrates are aligned with the interior topographies of human bodies, washed by the secret, sheltered tides of blood and breath. Branching hard corals resemble skeletal structures while feathery, spiny or meshed creatures suggest the networks of the nervous system or entangled neurons of the brain. I employ the complex morphologies of sea creatures to evoke frag-

Figure 8.5 Anne Taylor (2007). *Grip*. Etching and aquatint.

mented components of human anatomies and the unseen intricacies of human functioning, both physical and cultural. Conversely, soft coral formations are created from the magnified folds of interior organs, at once alien and familiar, enacting an inversion of interior and exterior, where the inner filigrees of body cavities morph into the marine environment. The scientific tools that allow such minute examinations of life appear in my work as arcane manipulations divorced from their practical functions. The play of the artificial and natural, the constructed and the organic, creates mutations of mimesis as a means of critique. I engage with themes of reproduction and fertility, linked to the creative impulse and conceived as an extension of generative natural forces (Figure 8.6).

The most polymorphous natural environments are the extremely fertile coral reefs, supporting such a diversity of species that new animals are still being discovered. Many corals can hybridise with similar species, displaying a malleability that makes their taxonomies difficult to resolve (Jones 2007, 47–48). The underwater world and its protean denizens demonstrate a heterogeneity and unpredictability that provides a potent metaphor for the qualities of flexibility and creativity

Figure 8.6 Anne Taylor (2008). *Respiration*. Oil on canvas.

required to change for the better humanity's environmental and cultural behaviours. As Elizabeth Grosz points out, 'The new is the generation of a productive monstrosity, the deformation and transformation of prevailing models and norms' (Grosz 2005, 30).

My work suggests qualities of fluidity, nurturing and touch, commonly associated with a feminine sensibility. It investigates darker aspects of women's experience that ecofeminists have linked to the exploitation of natural resources (Plumwood 1993; Merchant 1983). When hierarchies are established which place the human, and specifically the male human, at their peak, qualities associated with other entities are devalued and subordinated. The philosopher Emmanuel Levinas described femininity as 'a subterranean dimension of the tender', bearing 'a weight of non-significance' (Levinas 2005, 257), a negativity which reflects the exploitative thinking which still dominates the social, economic and political arenas, driving an unsustainable expenditure of resources and labour. I wish to suggest that 'feminine' values which are in fact equally valid for men or women, are as necessary to human flourishing in a shared world as the aggressive and acquisitive qualities currently valorised. Through the depiction of the mysterious and cryptic aspects of the natural world, as markers of a radical alterity, I aim to evoke the invisible lives of recondite beings as inhabiting worlds intimately and inextricably entwined with our own.

The goal is not, however, simply to present naturalistic images of cryptic animals, a task most effectively fulfilled by the skilful photographs and films found in the popular media. By integrating unusual forms of life into a figurative interaction with human concerns I aim to create an emotional and cognitive alliance, so that symbolisation initiates a meaningful awareness not only of raw being, but of a metamorphic interlinking. My work integrates such depictions into the affective and speculative realms of our awareness, instigating an imaginative and ethical engagement with a more comprehensive range of animal entities.

Conclusion: empathetic alignments

Our interactions with animals diverge according to historical conditions and the relationships with various creatures that these conditions encourage or allow. Since the Industrial Revolution technologically advanced societies have only rarely engaged with animals in mutual relationships, where the fine-grained knowledge of animal habits and habitats ensure survival, or the strength and abilities of animals are relied on to extend human capabilities. Direct, sensory experience of the animal world is reduced to a very limited range of artificial options. Often our most comprehensive interactions with animals are with pets, which are given honorary human status and valued for their individuality (Serpell 2005, 132). Complex technology masks instrumental uses of animals so that suffering may be minimised, but all too often it is merely disguised or hidden. The biological sciences provide us with increasingly detailed information about a very broad range of living entities, and subsequently these empirical data are translated into engaging presentations that allow us virtual entry into the lived environments of animals we may never physically encounter. Such popularised and homogenised representations of unfamiliar species widely disseminate a superficial knowledge but do not greatly disturb our anthropocentric habits. In contrast, animal images appearing in contemporary art, such as those I have discussed above, are integrated into critiques of human interactions with the natural world, prompting a complex, open-ended reception. Art instigates questioning, inviting viewers to actively engage with, and contribute to, the multi-layered meanings it offers.

Intuition is intrinsic to aesthetic experience, which generates an understanding that is neither precise nor final, but may induce new ways of imagining or at least may complicate conventional assumptions. Through its intuitive engagement with ideas and values immersed in environmental, social and historical contexts, contemporary art tests the boundaries of human and animal existence, loosening the bonds of anthropocentric thinking. A responsive engagement with aesthetic qualities, aligned with critical thinking embedded in social and environmental concerns, encourages an appreciation of the intrinsic worth of cryptic species, broadening our values beyond self-interested concerns and instilling respect for an extended range of animals and their environments.

Works cited

Albert R Mann Library (2009). Blaschka invertebrate models. [Online] Available: blaschkagallery.mannlib.cornell.edu/ [Accessed 18 February 2014].

Attfield R (2003). *Environmental ethics.* Cambridge, UK: Polity Press.

Baxter R (1998). *Bestiaries and their users in the middle ages.* Phoenix Mill, UK: Sutton.

Breidbach O (2005). Introduction: the most charming creatures. In E Haeckel, *Art forms from the ocean* (pp7–23). Munich: Prestel.

Breidbach O (2004). Brief instructions to viewing Haeckel's pictures. In M Ashdown (ed), *Art forms in nature* (pp9–18). Munich: Prestel.

Bryson N (1990). *Looking at the overlooked.* Cambridge, MA: Harvard University Press.

Calarco M (2008). *Zoographies.* New York: Columbia University Press.

Daston L & Mitman G (2005). The how and why of thinking with animals. In L Daston and G Mitman (eds), *Thinking with animals: new perspectives on anthropomorphism* (pp1–14). New York: Columbia University Press.

Foster H (1993). *Compulsive beauty.* Cambridge, Massachusetts: MIT Press.

Foucault M (1994). *The order of things: an archaeology of the human sciences.* New York: Vintage Books.

Gallagher E (2005a). Characters, myths and stories [Interview]. *Art 21.* [Online] Available: www.art21.org/texts/ellen-gallagher/

interview-ellen-gallagher-characters-myths-and-stories [Accessed 18 September 2013].

Gallagher E (2005b). eXelento and DeLuxe [Interview]. *Art 21*. [Online] Available: www.art21.org/texts/ellen-gallagher/ interview-ellen-gallagher-exelento-and-deluxe [Accessed 18 September 2013].

Gould SJ (1998). *Leonardo's mountain of clams and the diet of worms*. London: Jonathon Cape, Random House.

Grosz E (2005). *Time travels*. Durham and London: Duke University Press.

Grosz E (2004). *The nick of time*. Crow's Nest: Allen & Unwin.

Hunt S & Carter P (1999). *Terre Napoleon: Australia through French eyes*. Sydney: The Historic Houses Trust of NSW.

Jones S (2007). *Coral: a pessimist in paradise*. London: Little, Brown.

Kemp M (2007). *The human animal in Western art and science*. Chicago: The University of Chicago Press.

Levinas E (2005). *Totality and infinity*. Pittsburg: Duquesne University Press.

Malik A (2006). Patterning memory. *Wasafiri*, 21(3): 29–39.

Merchant C (1983). *The death of nature*. San Francisco: Harper & Row.

Plumwood V (1993). *Feminism and the mastery of nature*. London: Routledge.

Proctor R (2007). René Binet and the esquisses decoratives. In R Binet (ed.), *Rene Binet: from nature to form* (pp4–26). Munich: Prestel.

Rozwadowski HM (2005). *Fathoming the ocean: the discovery and exploration of the deep sea*. Cambridge, MA: The Belnap Press of Harvard University Press.

Serpell, J (2005). People in disguise: anthropomorphism and the human–pet relationship. In L Daston and G Mitman (eds), *Thinking with animals: new perspectives on anthropomorphism* (pp121–36). New York: Columbia University Press.

Smith B (1992). *Imagining the Pacific: in the wake of the Cook voyages*. Carlton, VIC: Melbourne University Press.

Wolfe C (ed.) (2003). *Zoontologies*. Minneapolis: University of Minnesota Press.

9

Linguistic anthropomorphism: *Timbuktu*, *The Whistler* and *The White Bone*

Sally Borrell

At the heart of ideas about human–animal relations is a negotiation between similarity and otherness, seemingly ever-shifting poles in the understanding of what it means to be alive in the world. If what constitutes the human as human is only difference from other species, then the nature of the human must remain contingent upon which species, which differences, and which understandings of difference come into consideration. Yet points of similarity have been equally important in understanding ourselves, if not other animals, as the rise of behavioural science amply illustrates. Conversely, as Graham Huggan and Helen Tiffin (2009, 154) find acknowledged in an American Veterinary Medical Association report, if extrapolation from animal to human is relevant, then so must the reverse be. This chapter examines fiction that explores what the worldviews of other species might be by according language to nonhuman animals. The discussion is specifically concerned with literature written for adults. It takes the examples of Paul Auster's *Timbuktu* (1999), Stephanie Johnson's *The Whistler* (1998) and Barbara Gowdy's *The White Bone* (1998), chosen because, though contemporaneous with one another, they illustrate a range of approaches to what I call linguistic anthropomorphism.

In a much-cited formula, 18th-century philosopher Jeremy Bentham (1990, 136) suggested that, in thinking about how humans should treat other species, 'the question is not, Can they *reason*? nor, Can they *talk*? but, Can they *suffer*?' Yet the question of animal language has ongoing significance. It has been one of the chief distinctions used to set humans apart from other species, having often been regarded as exclusively human. From this perspective, nonhuman animals' subjectivity or 'personhood' is regularly denied too. When it comes to companion animals, people often speak and behave as if animals are persons, but this is often regarded as something of an accepted joke, much like the infantilising description of pets as 'surrogate children' or 'fur babies.' Animal welfare policy, on the other hand, tends to reflect a presumption of difference until proven similar, in terms of reason, suffering, emotion, sentience, and other characteristics that go to make up concepts of subjectivity.

Fiction has an interesting role to play here because it can be used to lend language to animals. While this is by no means a prerequisite to the conclusion that animals possess subjectivity, it provides one means of exploring and speculating about what that subjectivity might entail. The resulting animal perspectives may come across as entirely imaginary – part of the temporary, fictional loan – but they also have the potential to stage real challenges to assumptions about the limitations of other species. Animals in *Timbuktu*, *The Whistler* and *The White Bone* are shown thinking or communicating in language while clearly remaining animals, thus opening up ideas about reason, suffering, subjectivity, and the limits of human understanding. What follows is an outline of the issues informing anthropomorphism and linguistic anthropomorphism in particular, and a close analysis of the sample literature comparing the canine perspectives of *Timbuktu* and *The Whistler* and then treating the unusual *The White Bone* as a separate case. On the basis of this analysis, I suggest that, far from limiting understanding of other species, linguistic anthropomorphism can provide an effective means of opening up questions about animal subjectivity.

Anthropomorphism

The greatest conceptual obstacle when imaginatively according language to animals is the tendency to read anthropomorphism as at best a fantasy and at worst a dangerous mistake. Etymologically, the word 'anthropomorphic' means 'human-shaped'. Examples of physically anthropomorphic animals, frequently upright, gendered, clothed

and bearing humanistic facial expressions commonly appear as alleg-
ories or within stories for children. An obvious instance is Kenneth
Grahame's *The wind in the willows* (1908) where Ratty, Mole, Toad
and Badger speak, wear clothes and are involved in adventures with
their boats, motor cars and the coming of the railway (species relations
become particularly confusing where they employ horse-drawn transport). In such narratives, animals have personhood, but it is imaginary,
and their readers are expected to understand or grow to understand
that it is imaginary. Huggan and Tiffin point out that,

> in 'reading' animals . . . we depend on generic indicators to dictate
> our appropriate responses. Talking animals belong in satire, fable,
> cartoons or children's stories . . . Anthropomorphism . . . would be
> inappropriate in accounts of eye-brain experiments on primates or
> octopus. (2009, 152)

Thus, while the physical morphing of animals can seem disturbing in
a novel like *The island of Doctor Moreau* (Wells 1996) where it results
from scientific experiment, Grahame's examples within the genre of
children's story can be safely interpreted as harmless, fantasy misrepresentation. Consequently, debates about anthropomorphism are not
often concerned with this variety. The question of misrepresentation,
or not, becomes less clear-cut and more interesting, however, where the
term is used in its more general sense to denote representations that
attribute human qualities to another animal. Most often, the term anthropomorphism is really used to mean *personification*, and whether
nonhuman animals are persons or have personhood is a very different
question from whether they physically resemble humans.

The various objections to anthropomorphism, and the sometimes
opposing values that inform them, are worth unpacking here in more
detail. Fundamentally, 'The basic logic to the anthropomorphism critique is that a category mistake is occurring because humans are radically different from animals' (Philo & Wilbert 2000, 19). Objections to
this mistake, whether motivated by anthropocentric concerns or the reverse, are founded on the values attached to the qualities of individual
species.

Anthropomorphism is sometimes criticised because it threatens
the construction of the human as unique, in respect of language, emo-

tion, reason, morality, or whichever other qualities we are seeking to claim as our exclusive province. A useful literary example of this response can be found in Yann Martel's *Life of Pi* (2001). Pi's father, who owns a zoo, believes that the most dangerous animal is '*Animalus anthropomorphicus*, the animal as seen through human eyes' (31). He tries to prevent his sons from anthropomorphising animals because it could place them in danger. Ostensibly for the same reason, Pi repeatedly attempts to dominate the tiger Richard Parker with whom he shares a lifeboat, at one point literally forcing him to jump through hoops (274). However, when Pi is later asked to retell his story without featuring Richard Parker, he describes a crisis of morale that suggests the assertion of his humanity as superiority is also psychologically important. Either way, it is for the protection of the human that Pi maintains that 'an animal is an animal, essentially and practically removed from us' (31).

The flipside of this objection is that to insist on similarity is an injustice to other animals. As Tom Tyler (2009, 15) puts it, 'by focusing on that which the animal shares with the human, we are in danger of missing all that is peculiar and proper to it' (see also Oerlemans 2007). From this perspective, anthropomorphism artificially restricts the field of understanding to that which is comprehensible in human terms. This conceptual limitation can in turn have very real consequences for animals and their treatment at the hands of humans. Abilities such as tracking by scent can be tested in tasks at which humans would fail abysmally and are reliant on canine assistance. Yet animal intelligence tests – tests which are often used to determine the acceptable treatment of a given species in, for instance, laboratory environments – are often designed to demonstrate capacity to perform human tasks or to think in human ways. They may demonstrate just that, but they demonstrate *only* that. In these terms, Tyler (2009, 16) is quite right to conclude that anthropomorphic thinking constitutes an 'anthropocentric failure of imagination'.

Such scruples on the subject of anthropomorphism are essential considerations. Animals' differences from humans are undeniably valuable, and to ignore these or to presume complete knowledge of animals is indeed likely to result, at best, in error. I would, however, also agree with Philo and Wilbert (2000, 19) that 'if the possibility is entertained that humans and animals may *not* be so completely different after all

. . . then the logical grounding for the charge of anthropomorphism becomes much more rickety'. A useful way of understanding the contradictory dynamics of anthropomorphism is as a spectrum. At one end are uses of anthropomorphism which serve human purposes, where animals are made to resemble humans not necessarily as stand-ins for them but in ways that complement established understandings of the human condition. At the other are uses that attempt to do the opposite, to interrogate and subvert the distinctiveness and privilege which humans have traditionally sought to accord themselves. As some recent animal studies scholarship demonstrates, anthropomorphism can offer effective opposition to anthropocentric thinking (see for instance Tiffin & Huggan 2009; Oerlemans 2007; Daston & Mitman 2005). Its value as a tool for such opposition is quite simply that within the context of fiction, which allows imaginative experimentation *without* presuming knowledge, it can be used to explore and raise possibilities.

Linguistic anthropomorphism: ventriloquism and translation

The specifically linguistic anthropomorphism that is my focus here also brings its own practical challenges: it necessitates putting words into animals' mouths without compromising their animality. The fact that nonhuman animals cannot, in human terms, speak for themselves does not necessarily dispel the problem of ventriloquism. JM Coetzee's work provides ample examples of reservations about this point. His novels often refuse to tell the stories of other racial groups, the most obvious instance being the symbolic muteness of Friday in *Foe* (Coetzee 1986), and Coetzee's subversion of Defoe's *Robinson Crusoe*. Similarly, *Disgrace* (Coetzee 1999) never gives the story of its African character Petrus; the protagonist David Lurie would like to hear it, but 'preferably not reduced to English. More and more he is convinced that English is an unfit medium for the truth of South Africa' (1999, 117).

Benita Parry (1994) suggests that Coetzee's approach repeats the exclusion of the silenced, but Gayatri Spivak argues that *Disgrace*'s refusal to 'give "voice"' to 'the subaltern' is the result of 'a politically fastidious awareness of the limits of its own power' (2002, 24). Given this sensitivity to the cultural implications of narrative, it is perhaps unsurprising that Coetzee displays similar restraint in response to species. His work has become increasingly concerned with animals in ways that refuse to allow readers to interpret them symbolically, yet, as Onno Oerlemans (2007) underscores, this is never achieved through ventriloquism; that is, through the author presuming to speak for animal characters. Coetzee's animals are 'silent others, usually visibly suffering at the hands of humans. Though they are closely watched by the novels' protagonists, they are largely inscrutable' (Oerlemans 2007, 185). Instead, Coetzee offers a 'pervasive and implicit deconstruction of the difference between animal and human' that 'dissolves notions of our own superiority' (2007, 185). Coetzee's work interrogates human attitudes to animals rather than speculating about the reverse, again perhaps out of respect for the limits to what can be assumed about other perspectives.

Timbuktu, *The Whistler* and *The White Bone*, by contrast, do not avoid ventriloquism. Instead, they mediate it by employing 'translation' (in one sense or another) as a way of bridging rather than denying the space between the reader and the animal characters. *Timbuktu* and *The White Bone* achieve this via a third-person narrative perspective that serves as a kind of intermediary, while the first-person narrative of *The Whistler* is presented as the result of special interpreting software. In any translation between languages, it is necessary to arrive at a balance between literal and communicative approaches; that is, between a word-for-word translation and one that best expresses and conveys essential meaning. The appropriate mix of these in each case will depend on the genre of the source material and the intended purpose of the translation. Within the linguistic anthropomorphism of these novels, the space between these literal and communicative considerations equates to space for differences in animal and human communication, while rendering animal perspectives accessible. Notably, however, the canine protagonists of *Timbuktu* and *The Whistler* need no such assistance: they understand human speech without much difficulty, allowing these novels not only to speculate about dog perspectives but also to comment on humans.

Canine perspectives: *Timbuktu* and *The Whistler*

At the outset of *Timbuktu*, the canine protagonist Mr Bones knows only his life with Willy, a would-be poet living on and off the streets. Willy is dying, and the story follows Mr Bones as he reflects on their life together and struggles to find a new place within human society. Canine and human languages are presented as distinct. Mr Bones understands 'Ingloosh' as a 'second language', helped by 'the advantage of being blessed with a master who did not treat him as an inferior' (6). This language is

> quite different from the one his mother had taught him, but even though his pronunciation left something to be desired, he had thoroughly mastered the ins and outs of its syntax and grammar. (6)

Mr Bones therefore has the advantage of knowing what humans say (in English – he does not, for instance, understand Chinese [108]). His thoughts are also rendered in English, whether or not he is actually thinking in this language, offering the reader accessible third-person insight into his world.

Somewhat paradoxically, however, Auster's approach also maintains a degree of authorial distance from the very proximity this permits. Jutta Ittner (2006, 182–83) sees *Timbuktu* as an example of 'a paradigm shift from the traditional anthropomorphic view ... to [one] that views animals as a separate and unknowable identity'. While we can only conceive of other animals from within human consciousness, which is inevitably reflected back to us in such texts, Ittner (2006, 183) writes that 'The new anthropomorphic approach acknowledges this impasse and integrates it into its inquiry on animal alterity'. However, the presentation of Mr Bones is also filtered through a veil of humour. Ittner comments that the narrative stance is difficult to locate 'because the tone ... fluctuates between amusement ... and ridicule, condescension and even cynicism', juxtaposed with a real interest in and sympathy for Mr Bones (2006, 184–85). This comic tone could be interpreted as self-consciousness about anthropomorphism, implying that Auster is not quite serious and so exempting him from accusations of sentiment. Moreover, within the novel's acknowledgement of unknowable animal alterity, problematic echoes of human cultural 'othering' surface

to create tension between maintaining and challenging traditional hierarchies.

The rendering of Mr Bones's thoughts in language, together with
occasional commentary by the third-person narrator, gives him complete and complex personhood. Mr Bones possesses both profound
emotions and logical reasoning, and he reflects upon human nature
and forms individual attachments. He displays the loyalty and devotion
considered characteristic of dogs, but he also understands the predicament he is in and knows that he must ingratiate himself into the society
of different humans to survive. He learns who to trust and shows an
understanding of who has decision-making power and how to impress
different people. When he comes across a somewhat stereotypical suburban family, he exhibits reason and self-control in order to conform to
human preferences and make a good impression (140).

Species distinctions, however, are not denied in the text. One of the
greatest discrepancies is that Willy cannot understand dog language in
the way that Mr Bones can understand English. His resulting mystifica-
tion is represented in terms reminiscent of colonial cultural contact: for
Willy, life with a dog is 'like learning how to speak a new language . . .
like stumbling onto a long-lost tribe of primitive men and having to figure out their impenetrable mores and customs' (37). Yet, also like many
colonisers, learning to speak Mr Bones's language is not something that
Willy attempts at all. This interesting imbalance could be interpreted as
presenting animals as more intuitive, or as an acknowledgement of the
real limitations to human knowledge of animals.

Willy does try to understand Mr Bones's experience of smell, but
is unsuccessful. He tries to produce 'olfactory art' (41) for dogs, since
Mr Bones might aspire to 'things not necessarily related to the needs
and urgencies of his body, but spiritual things, artistic things, the im-
material hungers of the soul' (40). This attitude is refreshing, but Mr
Bones enjoys Willy's artworks simply because 'dogs enjoyed smelling
whatever they were given to smell . . . There is . . . nothing to separate
the high from the low' (44). There is, in other words, no such thing as
art as a distinct experience for dogs. The limitations of Willy's knowledge are again apparent, but like the earlier use of the word 'primitive',
traditional understandings of human–animal distinctions appear too:
Mr Bones's experience of life, according to the narrator, excludes the

separation of mind and body so often used to set humans above other species.

The novel's greatest challenge to such hierarchies occurs in relation to Mr Bones's morality and perception.

> Willy had judged him to be wholly and incorruptibly good. It wasn't just that he knew Mr Bones had a soul . . . the more he saw of it, the more refinement and nobility of spirit he found there. Was Mr Bones an angel trapped in the flesh of a dog? Willy thought so. (35)

Mr Bones possesses an instinct for truth and at times is even prescient. Willy 'had told so many stories, had spoken in so many different voices, had spoken out of so many sides of his mouth at once that Mr Bones had no idea what to believe anymore' (12). It is implied that Mr Bones's perspective is more reliable, and Ittner suggests that Auster 'elevates his dog to the role of a messenger from a pristine world that humans have lost' (193). Mr Bones also has a prophetic dream about Willy's death, and when it begins to come true, he flees. Although they are separated, the moment of Willy's death has a physical impact on him (97). These moments hint at heightened powers, extending a word play of Willy's by suggesting ways in which dogliness may be close to godliness.

Ultimately, however, final rewards are understood in human terms. Even in Mr Bones's imaginings of Timbuktu – heaven – his hope is that 'dogs would be able to speak man's language and converse with him as an equal' (50, 181). His wish does not involve a heavenly Willy speaking his first, canine language; when Mr Bones effectively commits suicide (in a final game of 'dodge-the-car'), it is in order to reach this Timbuktu where he can speak human language.

Overall, *Timbuktu* uses linguistic anthropomorphism to accord personhood to a dog and to explore human–animal relations from his perspective without attributing any subversive tendencies to him. Mr Bones is largely accepting of his lot, and it is for the reader to judge the circumstances in which he finds himself. Although he is obviously at the mercy of humans, what Mr Bones himself wants most is a way to overcome the language barriers that keep him from communicating with humans fully. By contrast, the title character of Stephanie Johnson's *The Whistler* achieves exactly that, yet mixes feelings of individual loyalty with more general resentment of human–animal power relations.

Set in a futuristic, dystopian Sydney, *The Whistler* is narrated by Smooch, a legless lapdog living with two humans: Verity, a sex worker and her deformed, gilled son Vernon. When Verity disappears, Smooch and Vernon venture out into a dystopian world dominated by the Corporate Principle, where power is in the hands of the few. Even more than Mr Bones, Smooch understands and thinks in human words. He displays not only reason and emotion, but an understanding of irony, eloquence and the intricacies of human language: 'I not only try to keep my brainwaves seemly, but my grammar correct' (viii). In this case, Smooch's language skills are part of the science fiction of the novel; he can also remember his past lives. Vernon has developed an electrode that lets Smooch narrate his previous incarnations to a computer, where they are converted into text which punctuates the main narrative. He has been sometimes male and sometimes female, but has always had a human Lady companion; including the Virgin Mary, Mary Queen of Scots, Kupe's wife Hine Te Aparangi, and a member of a futuristic feminist biosphere. This means that Smooch, perhaps even more than Mr Bones, provides a window into the human world and human nature.

As in *Timbuktu*, *The Whistler* uses humour to manage the juxtaposition of Smooch's intellectual eloquence with the physicality more typically understood to define animal being. He repeatedly expresses a wry sense of shame at a body bred without legs. '*Lapdog*. Oh how clearly I can see, with hindsight, how that . . . insignificant word precipitated my final nemesis' (xi). Linguistic anthropomorphism is mediated by the same irony that Auster employs, as well as by the science fiction which here serves as a further 'excuse' for it. But even as this facilitates the readers' acceptance of the approach, the limits to human understandings of animal being are a recurring theme. Just as Mr Bones hates being renamed Sparky, Smooch considers his name the worst he has ever had. This implies that their subjectivity is informed by a sense of their own dignity which humans do not always acknowledge. Meanwhile, Verity does not believe Vernon's assertion that Smooch smiles until she sees it for herself (25, 54), and initially doubts Smooch could have an imagination (24). In fact, Smooch reflects that he is constantly making creative decisions about his narrative and would prefer them to wait for his work in progress to be completed (51). While Smooch comes across as intellectually equal or even superior to humans, humans repeatedly underestimate him.

Smooch's story also includes a more serious interrogation of human–animal relations, with particular attention paid to the suffering caused by selective breeding. Smooch is in the opposite position to Mr Bones: as a mongrel, Mr Bones struggles to find humans who will accept him, whereas Smooch has been so specifically bred to suit that he is physically dependent on humans.

> It was, I recall, in 1645 that the word 'lapdog' officially entered the English language . . . Once the fashion reached the middle classes, there was an increased demand for my kind. Breeders were only too happy to oblige. (x)

This entirely anthropocentric shaping of dogs has continued to the point that Smooch has no legs, only vestigial stumps or paddles. In addition,

> Just as spaniels suffer ear infections, borzois leg cramps, beagles asthma and Afghans microcephaly, Whistlers suffer kidney failure. It is our sedentary lifestyles that leave us open to it, our grinding, stationary organs. (x–xi)

It also becomes clear that humans have become the subjects of breeding programs and genetic engineering too. Healthy children are rare and valuable. Smooch and Vernon encounter a band of parentless children living in an old hospital. Smooch realises they bear the mark of the Breeders, like him. They are the imperfect, unwanted results of genetic engineering to produce perfect children: 'seconds, fire-sale items, damaged goods' (101). These children have deformities like Smooch and Vernon, but unlike them, they have been rejected. Doubtless informed by Johnson's own history of disability (Johnson 1998a), the novel explores the experience of being bound by and judged according to a particular bodily form. This experience is heightened for Smooch because he remembers earlier, active lives.

The apparent deformities of Smooch, Vernon and Liban, a girl from the hospital, turn out to be adaptations equipping them for a new life in the sea. Smooch discovers that he can swim like a turtle. Vernon, who increasingly struggles to breathe on land, uses his gills to breathe underwater, and Liban's oversized webbed feet serve as flippers. They are not the only characters like this: the family's neighbours have a Whistler with a beaver's tail, and a baby daughter with gills. Despite the Breeders' interventions, the physical characteristics of different animals are naturally combining to cross species boundaries in ways that constitute not mutations but advantages. The results are liberating both physically, and for Smooch, existentially as well. It is as if he has reached a final nirvana at the end of all his reincarnations: for all his attachment to humans, he rejoices that finally, 'the lapdog will be free of the lap, the confines of his Ladies' existences' (235).

The Whistler, like *Timbuktu*, is another experiment with narrative perspective. However, the seriousness of the issues of breeding or genetic modification and bodily otherness in *The Whistler* is stressed by the first-person perspective. The result is a far stronger sense of exploitation and confinement that opens up different and more difficult questions about the ethical limits to human power over life, or the lack of such limits. Again, however, the closeness of human–canine relations facilitates the evocation of the animal's perspective. As Ittner (2006, 183) comments,

> It is hardly surprising that the most abundant literary animal is *canis familiaris* . . . Unlike any of the other species that have served as pets, dogs form the close, merging relationships with their humans that satisfy our need to be mirrored.

It need not follow that this mirroring be flattering. In both *Timbuktu* and *The Whistler*, it is an instructive self-reflexive exercise to imagine how animals might experience humans. My final literary example, however, takes the opposite approach. I turn now to an exploration of what can be achieved by shifting the focus *away* from human–animal relations as Barbara Gowdy does in *The White Bone*, where humans only ever figure on the margins.

Elephant visions: *The White Bone*

Barbara Gowdy's *The White Bone* imagines an intricate world of wild animals, exploring the impact of hunting and climate change on endangered African elephants and making an urgent call for their protection. The elephant protagonist Mud, her herd and the other herds in the area are struggling to survive, but believe that if they can find a mythical White Bone, it will show them the way to a wildlife reserve they call the Safe Place. Humans seldom appear in the novel, but when they do – with the exception of those at the Safe Place – they constitute one of the two greatest threats to the elephants. As Huggan and Tiffin (2009, 150) write, the novel reverses 'the human/animal dichotomy in relation to savagery and civilisation'. Although the elephants' language is rendered in English, it is not a human language in this novel: indeed, humans and snakes are the only species with which the elephants cannot communicate. The language is also modified to 'elephantise' it. An accompanying glossary lists terminology such as 'feast tree' (acacia), 'hindlegger' (human), 'tall time' (dawn or dusk) and 'rogue's web' (wire fence).

Novelist David Mitchell (2010, 559) has observed that in writing historical fiction, one problem he faces is that while modern language is inappropriate, 'a "correct" translation into Smollett's English . . . smacks . . . of Blackadder, and only a masochist could stomach 500 pages'. Therefore, he writes:

> To a degree, the historical novelist must create a sort of dialect – I call it 'Bygonese' – which is inaccurate but plausible. Like a coat of antique-effect varnish . . . it is both synthetic and the least-worst solution. (2010, 559)

Gowdy has similarly created a dialect for her elephant characters, made out of human language but with a 'varnish' of otherness designed to evoke an elephant's perspective. Just as Mitchell's characters speak 'Bygonese', hers speak 'Elephantese'.

This language reveals a complex culture, both like and yet unlike those of humans. The elephants live in matriarchal herds where both cows and bulls are known as 'she-ones' (a term Gowdy compares to 'mankind'). They possess religion, superstition, mysticism and even song and visual art. They converse both casually and in a 'formal timbre', '[a] respectful form of address characterised by exaggerated enunciation' (xiv). Like Mr Bones and Smooch, the elephants possess skills that humans do not. In Mud's herd, her friend Date Bed is a 'mind talker'; she can communicate telepathically with elephants and members of other species who provide valuable information about survival. Mud is a visionary, meaning that she can see 'both the future and the distant present' (xviii). These skills occur in only one cow at a time, and never in males, though the bull Tall Time is something of a clerical figure, studying 'links' or omens. All the elephants also have an advanced capacity for memory, 'preserved inside them as a perfect and instantly retrievable moment' (1).

Indeed Gowdy goes beyond anthropomorphism. Oerlermans (2007) reads the culture of Gowdy's elephants according to three connected categories. The first describes behaviour observed in real elephants, the second involves anthropomorphism based on that behaviour, and the third extrapolates from this to assume complex consciousness. Where Gowdy takes this a step further into fantasy, Oerlemans makes a convincing argument that in doing so, she 'sets out not just to suspend disbelief . . . but also to provoke it' (2007, 194). By this means, he suggests, the novel 'goes some way toward deconstructing the very concept of anthropomorphism, forcing us to question which aspects of being and consciousness are, after all, purely human' (2007, 195). What is particularly productive in Gowdy's foray into implausibility is that it implies that the subjectivity of other animals may go beyond what seems probable or even comprehensible to humans. In this way, where *The Whistler* emphasises humans' underestimation of other animals, *The White Bone* implies that animal experience may be so far from human experience that we can scarcely credit it at all.

The elephants' heightened capacities, particularly for memory and emotion, also mean that their vulnerability may fall outside human comprehension. They are prone to a kind of shell-shock where they become lost in memories. After a massacre by hunters, Mud becomes caught up this way.

[B]y the time she understands that the screams come from memory, she is reliving the slaughter. She trumpets and runs in circles. At the part where she climbed up and slid down the bank, she climbs up and slides down the rocks and scrapes her leg badly enough to arouse her to the present . . . she sinks onto her right side and weeps, for how long she has no idea. (94)

The elephants are also at risk from their interdependence on one another's unique abilities. During the attack on Mud's herd, Date Bed flees in panic and becomes separated from the others, but until she dies, no-one else in the group can become a mind-talker and they are deprived of essential lines of communication with other species. Desperate for help, they are forced to bargain with a cheetah – who wants first-born calves in return for information – through a comparatively 'rudimentary' sign language (249–50). Date Bed eventually finds the White Bone only to die just before the herd reaches her, and it is only because Mud now inherits Date Bed's telepathic ability that the nearby mongooses can tell her what direction to take. The elephants' greater sensitivities thus exacerbate their plight. Significantly, Gowdy never reveals whether the elephants find their destination. Instead, their final position reflects that of real elephants in contemporary Africa: in great danger of dying (out) before they reach safety.

The White Bone personalises the dangers of environmental destruction and slaughter by conveying their impact on a complex and fragile culture. The invention of a special 'Elephantese' and telepathy invites speculation not only about animal personhood but about the potentially very different scope of animal being. Far from coming across as anthropocentric in its anthropomorphism, Gowdy's leap of faith insists that whatever other animals' lives are like, the extent to which they are comprehensible by humans is likely to be profoundly limited, and that consequently we underestimate them. I would by no means suggest that linguistic anthropomorphism always does this; on the contrary, it may just as often have the opposite effect. However, *The White Bone* illustrates the extent of its potential as a device for undoing the centrality of the human and interrogating the limits to our understanding of other species and our impact upon them.

Language and subjectivity

The uses of linguistic anthropomorphism in these three texts are revealing in both their differences and their similarities. In terms of the purpose of linguistic anthropomorphism, two different factors come into play in these texts. Firstly, the novelty of 'talking' animals in fiction for adults provides scope for narrative experiments. It is refreshing yet challenging to write or read from the perspective of another species and to use that perspective as a metaphor for or mirror of human being. A second, less traditional purpose of linguistic anthropomorphism is offer an insight into what animals might experience and to foster a sense of urgency about issues affecting them, thereby fostering a desire for change or action. The novels I have discussed occupy different positions on this anthropomorphic spectrum according to their aims. Auster's *Timbuktu* uses animal language to give a canine perspective on human–animal relations, and *The Whistler* uses it to draw these into question, while *The White Bone* underscores the limits to human knowledge as part of an overt promotion of ecological responsibility.

The commonalities between these texts also offer insights into the workings of linguistic anthropomorphism. Both Auster and Johnson make use of humour, often at the expense of their canine protagonists, in ways that provide comic relief but also create some ironic distance from the experimentation with anthropomorphism. In all three novels, the exploration of otherness includes the idea of other-than-human perception or sensitivity, such as in Mr Bones' premonitory dream, Smooch's memory of his past lives, and the elephants' powers of memory and clairvoyance. These recurring features of humour and fantasy show the importance of creative licence in asking readers to countenance linguistic anthropomorphism.

Conclusion

Although anthropomorphism is subject to critique as undervaluing the distinctions between humans and other species, fiction escapes the different-until-proven-similar restriction that often applies to speculation about animals. In their discussion of anthropomorphism, Philo and Wilbert (2009, 19) recommend

a measured, hesitant and reflected-upon form of anthropomorphism
. . . which would allow the possibility of insights to be produced
from considering some nonhumans in some situations as if they
could perceive, feel, emote, make decisions and perhaps even 'reason'
something like a human being.

The loan of human language to animals in the novels discussed here
fosters an approach like that suggested by Philo and Wilbert: readers
are encouraged to speculate about animal subjectivity and personhood
and to consider the likelihood of unproven commonalities. This use
of linguistic anthropomorphism provides further support for the view
that anthropomorphic thinking need not always be regarded as an error; instead, it can be an important perspective in the conceptualisation
of human–animal relations.

Works cited

Auster P (1999). *Timbuktu*. London: Faber.
Bentham J (1990 [1789]). An introduction to the principles of morals and
    legislation. Extracts in A Linzey & PC Clarke (eds), *Political theory and
    animal rights* (pp135–36). London and Winchester: Pluto Press.
Coetzee JM (1999). *Disgrace*. London: Random House.
Coetzee JM (1986). *Foe*. New York: Viking Press.
Daston L & Mitman G (2005). *Thinking with animals: new perspectives on
    anthropomorphism*. New York and Chichester: Columbia University Press.
Grahame K (1908). *The wind in the willows*. London: Methuen.
Gowdy B (1998). *The White Bone*. New York: Picador.
Huggan G & Tiffin H (2009). *Postcolonial ecocriticism*. London and New York:
    Routledge.
Ittner J (2006). Part spaniel, part canine puzzle: anthropomorphism in Woolf's
    Flush and Auster's *Timbuktu*. *Mosaic*, 39(4): 181–97.
Johnson S (1998a). Interview by Murray Waldren. *The Weekend Australian*.
    [Online] Available: www.users.tpg.com.au/waldrenm/johnson.html
    [Accessed 5 August 2012].
Johnson S (1998b). *The Whistler*. Auckland: Vintage.
Martel Y (2001). *Life of Pi*. Edinburgh: Canongate Books Ltd.
Mitchell D (2010). *The thousand autumns of Jacob de Zoet*. London: Spectre.
Oerlemans O (2007). A defense of anthropomorphism: comparing Coetzee and
    Gowdy. *Mosaic*, 40(1): 181–96.

Parry B (1994). Speech and silence in the fictions of JM Coetzee. *New formations*, 21: 1–20.

Philo C & Wilbert C (eds) (2000). *Animal spaces, beastly places: new geographies of human–animal relations*. London: Routledge.

Spivak GC (2002). Ethics and politics in Tagore, Coetzee, and certain scenes of teaching. *Diacritics*, 32(3–4): 17–31.

Tyler T (2009). If horses had hands . . . In T Tyler & M Rossini (eds), *Animal encounters* (pp13–26). Leiden: Brill.

Wells HG (1996 [1896]). *The island of Doctor Moreau*. Edited by Leon Stover. Jefferson, NC: McFarland.

Part III
Animal and human welfare

10

TNR (trap-neuter-return): is it a solution for the management of feral cats in Australia?

Mandy Paterson

Unowned, free-roaming cats exist in most urban and rural environments around the world. Cats are widespread colonisers and can establish colonies in many, quite different ecosystems (Tennent et al. 2009). They are also prolific breeders, becoming sexually active from five to six months of age and producing one to six kittens 1.4 times per year (Nutter et al. 2004a). Worldwide their numbers are difficult to accurately assess but are most likely in the billions. It has been estimated that there are between 12 and 19 million feral cats in Australia (Jongman & Karlen 2006; West 2008). Such cats pose a number of serious threats to wildlife through predation, disease spread and competition; to public health through zoonosis; to public comfort through noisy displays; and to the health of owned cats through fighting and disease transmission (Robertson 2008; Schmidt et al. 2007). There are also concerns about the welfare of the cats themselves due to disease, fighting, low-nutrition diets, lack of veterinary care, and inhumane eradication attempts (Robertson 2008; Mendes-de-Almeida et al. 2007).

There is general agreement amongst government agencies, conservationists and animal welfare organisations that feral cat populations need to be managed; however, there is debate about how this management should be achieved, for example, whether control of feral cats using lethal methods is acceptable. Furthermore, with the wider public cats in many ways represent a high-profile species (Witmer et al. 2005) and discussion of any control technique has the power to evoke strong emotions.

Trap-neuter-return (TNR) is one approach to feral cat management which is gaining popularity in many countries, particularly in the US (Denny & Dickman 2010). It is especially popular in urban areas (Finkler et al. 2011a). TNR involves capturing feral cats, sterilising them and returning them to the place where they were found. The colonies are then managed through regular feeding by volunteers.

This chapter discusses feral cats in Australia, touches on the control measures currently available and examines TNR as a viable control option. It examines the ethical, conservation, animal welfare and legal issues with respect to feral cats, feral cat control and TNR. The chapter concludes that the control of feral cats is complex, that TNR could work in specific well-defined areas but in general is not a solution to the problem in Australia.

What's in a name?

The language used to describe unowned, free-roaming cats varies markedly. Some researchers use the term 'feral' to describe cats that are not socialised and therefore not able to be rehomed (Slater 2005; 2002). This purely behavioural definition, while useful when considering rehoming options, ignores the fact that cats can be sociable yet still be free-roaming and unowned. Moodie (1995) and Farnsworth et al. (2011) argue for a division into domestic (or companion), feral and stray, classifying feral as those cats completely independent of humans and strays as those relying to some extent on humans. This definition of 'feral' relies on the extent of human input but this can be difficult to assess. There is also strong evidence that cats move freely between urban sites where humans may help with feeding and more distant sites where they completely rely on predation (Guttilla & Stapp 2010; Denny 2005). Other researchers use the term 'free-roaming' and include all cats, whether owned or not, that are not confined and can therefore impact on ecosystems (Finkler et al. 2011a). This definition is based on level of confinement and not on ownership or socialisation status.

Cats that are part of the feral population may come from many sources. They may be domestic cats that have strayed or been lost or abandoned, cats born in the wild, or cats deliberately introduced to control vermin (Robertson 2008). Irrespective of source, how friendly they are, and whether they interact with humans or are dependent to some extent on humans for food, they roam freely. Their impact on the environment in terms of predation and disease spread, their public nuisance and concerns for their welfare, remain. Therefore, I argue, all such cats should be considered together and the term 'feral' should include all unowned, free-roaming cats irrespective of socialisation status or source.

Attitudes to feral cats vary widely within the community (Lord 2008) with 'ill-informed and soft-hearted urban dwellers' (van Heezik 2010, 153) at one extreme and ornithologists and wildlife biologists at the other (Loyd & Miller 2010a). One study in the US (Loyd & Hernandez 2012) found that more residents had positive attitudes to feral cats than negative, even though over a half of them were aware of the negative impact of cats on native species. Attitudes are also influenced by socioeconomic status, with cats living in low-income neighbourhoods less likely to be neutered and more likely to be pregnant and roaming free, compared to cats in high-income neighbourhoods (Finkler et al. 2011b).

In sum, the terminology used to describe cats can influence people's perception of cats, and importantly, their perceptions about the acceptability of control measures including TNR (Wilken 2012; Farnsworth et al. 2011; Lord 2008; Loyd & Miller 2010a; 2010b).

Feral cat control

Various methods (lethal and non-lethal) have been suggested to control feral cats including poison; trap and euthanase; trap, neuter and return to the home colony (TNR); trap, neuter, return and rehome kittens (Loyd & DeVore 2010); and trap, neuter and place in a cat sanctuary (Loyd & Hernandez 2012). In a standard TNR program, colonies are managed and management involves daily feeding of the cats as well as the trapping and neutering of new cats to the colony. There are also control approaches which focus on preventing recruitment to colonies such as educating cat owners to neuter their cats, discouraging unwanted behaviours such as the abandoning of cats or allowing them to breed, and enacting laws which promote responsible cat ownership (Finkler & Terkel 2012). Another consideration is the cost of control measures in terms of resources, time and ongoing effort (Jongman & Karlen 2006; Nutter et al. 2004b).

According to the Australian Government's National Consultative Committee on Animal Welfare (2008), any control program should protect the welfare of cats, reduce the impact of cats on wildlife, reduce the public nuisance of cats, recognise the value of cats to our community and educate the community. Any control program must consider its impact on non-target species (Andersen 2007). Accurate and appropriate cat-number assessment methods must also be available to assess control programs (Bengsen et al. 2011).

The aim of feral cat control

Different groups in the community have different desired outcomes for feral cat control and these differences add to the complexity of any discussion about control methods. The desired outcomes also influence assessment measures and decisions about whether a control program has been successful.

Conservationists in general aim for the reduction in the number, or the eradication, of cats in a particular ecosystem to reduce or eliminate their impact (Denny & Dickman 2010). Even if this outcome is agreed upon there may be disagreement about the acceptable timeframe in which this reduction or eradication should occur. Animal welfare proponents, conversely, tend to desire the existing population to be healthy such that they can live out their lives in a state of improved health and wellbeing (Robertson 2008; Winograd 2007).

It is immediately obvious that these two aims are difficult to reconcile. It is also true that conservation scientists and TNR advocates rarely meet or discuss feral cat control (Lepczyk et al. 2010; Longcore et al. 2009). This polarised state of affairs is unsatisfactory and I believe the wide range of stakeholders interested in feral cats and feral cat control must become involved in a scientifically informed discussion of the issue.

Ethical considerations

A fundamental ethical question is the extent of our responsibility towards feral cats in Australia considering that they are a species we introduced, and our responsibility to indigenous species and the environment. This question is underpinned by our attitude towards these species, and our attitude, I argue, is influenced by the value we place on them. Research suggests that increasing education levels and affluence, urbanisation and a more mobile society have led to a shift in societal values (Manfredo et al. 2003). There is a shift from materialist values including physical and economic security, to post-materialist values such as quality of life and self-expression (Teel et al. 2007) and the desire for a more humane society in general.

Paradoxically, this desire for a more humane society may result in either support for lethal methods of control of feral cats due to their negative impact on ecosystems or the opposite: support for the rights of feral cats to exist in our environment (Loyd & Miller 2010a). What influences the support is whether feral cats are considered to be wildlife (Wilken 2012) or if an introduced species (cats) have the same rights as native wildlife (Jongman & Karlen 2006). Wild animals are considered to benefit from living wild and humans tend not to deliberately interfere in their lives. Should this be true for feral cats as well?

TNR has its own ethical issues such as the acceptability of interfering with a free-living animal and then returning it, altered, to its home range. Jessup (2004), for example, argues that neutering may change the animals' success and even their welfare in the wild since neutered animals are thought to be lower down the feline hierarchy than entire animals.

TNR is an important element in the 'no kill' movement in the US (Winograd 2007). Feral cats that are not socialised adequately cannot be rehomed so TNR is the logical solution: they are returned to their colony and do not affect the no-kill statistics. However, is it ethical to return an animal to the wild and perhaps an uncertain future to maintain a no-kill status?

Conservation considerations

There is general agreement that cats have a significant negative impact on ecosystems (Loyd & DeVore 2010; Witmer et al. 2005), including on islands (Campbell et al. 2011) and specifically to birds (Dauphine & Cooper 2009). Feral cats represent a threat to over 110 species in Australia, more than the threat posted to Australian native species by any other exotic animal or plant (Coutts-Smith et al. 2007). This threat is usually through predation but can include through the spread of disease and competition (Denny & Dickman 2010). This threat is recognised by the government under its *Environment Protection and Biodiversity Conservation Act 1999* (Cwlth) which classifies the proliferation of feral cats as a 'key threatening process'.

Cats predate on small- to medium-sized mammals, birds, reptiles, amphibians and insects, and cats are often at densities ten to 100 times higher than other similar-sized predators (Liberg et al. 2000). A study in suburban bushland in Sydney found that the presence of cats decreased the richness of bird species (Dickman 2007). A study in suburban Perth found, however, that cat density was not correlated with bird-species richness, rather distance to nearby bushland, housing density and size of nearby bushland were (Grayson et al. 2007).

Cats are on the list of the 100 worst invasive species globally (Lowe et al. 2000). Even neutering does not appear to necessarily curb their hunting behaviour (Jongman & Karlen 2006). Some research suggests the home range of neutered cats is reduced (Hill 2006) whereas others have found no significant reduction (Guttilla & Stapp 2010). Neutered cats in TNR programs tend to live longer than intact cats, and are healthier and stronger (Robertson 2008; Schmidt et al. 2007), therefore they continue to have significant negative impacts on native wildlife (Guttilla & Stapp 2010; Levy & Crawford 2004). In addition, Schmidt et al. (2007) found that managed colonies lead to higher densities of free-roaming cats at that particular location through immigration, therefore predation increases locally.

A positive aspect of the presence of cats is their role in controlling other pest species such as mice and rats. Studies on islands have shown that if cat numbers are reduced, rat and rabbit populations skyrocket (Robertson 2008). Controlling cat populations is clearly not the only answer for positive conservation outcomes; all species impacting on native animals must be controlled.

Grayson and Calver (2004) argue that the precautionary principle should be adopted when considering the effect of cats on wildlife populations. The precautionary principle states that when there are threats with potentially serious consequences, the lack of full scientific knowledge should not prevent measures being taken to prevent these threats. That is, there is a need for action despite uncertainty.

Animal welfare considerations

Animal welfare concerns focus on individual animals. Welfare considerations should, of course, not only include the welfare of feral cats but also the welfare of the animals upon which they prey. Feral cats are generally known to kill and maim native animals (Coutts-Smith et al. 2007) but little systematic research appears to exist looking at the welfare of native animals predated on by feral cats.

There is debate about the benefits of TNR to feral cats. Robertson (2008) reports their health and body score index improve, and their life expectancy increases. However, other studies report no improvement in welfare (Jessup 2004). Jessup (2004) even goes as far describing TNR as trap-neuter-re-abandon. Certainly, pathogens tend to disseminate throughout a colony, and colony cats suffer from fleas and other parasites (Mendes-de-Almeida et al. 2007).

The process of capture, handling, surgery and transportation of feral cats can be distressing and have welfare implications, but these can be successfully managed (Looney et al. 2008). A reasonable number of trapped cats will be pregnant. Scott et al. (2002) reported a pregnancy rate as high as 47 percent during the breeding season, whereas Wallace and Levy (2006) found a pregnancy rate of 15.9 percent while undertaking a TNR program. Neutering for pregnant cats is major surgery with increased surgical and welfare risks.

Some have argued that managed colonies can become animal hoarding by another name (Lepczyk et al. 2010), resulting in negative welfare as carers exceed their capacity to care. In addition, there is some evidence that feral cats may be reservoirs of feline immunodeficiency virus, feline leukaemia virus and *Toxoplasma gondii* (Spada et al. 2012;

Little 2011; Levy et al. 2006). These diseases can have a welfare impact on owned cats and in some cases, on native species.

Welfare considerations, therefore, are clearly complex and multifaceted. Native animals suffer negative welfare from cat predation, but if cats are removed they could suffer negative welfare from other causes such as from other predators or overabundance. The cats themselves may enjoy improved welfare if included in a TNR program, or they may not; the research is inconclusive on this point and may depend on how welfare is measured.

Legal considerations

Each state in Australia has its own laws with respect to wildlife and animal welfare. A thorough discussion of all relevant state laws in Australia is beyond the scope of this chapter (see Denny & Dickman 2010, 9–11, for a discussion on the legal status of cats in Australian states). In general, many state laws do not support the implementation of TNR programs since they deem it illegal to abandon an animal or release a pest species. For example, in Queensland it is unlawful to release a declared pest animal back into the wild under the *Land Protection (Pest and Stock Route Management) Act 2002*; in Victoria it is an offence to abandon a cat (or dog) under section 33 of the *Domestic Animal Act 1994*, and it is also an offence under Section 9.1h of the *Prevention of Cruelty to Animals Act 1986*.

Does TNR work?

Whether or not TNR is deemed to have worked depends on the desired outcome. One can generally assume that conservationists wish to preserve endangered species and populations and reduce (or eliminate) predation. This is usually assessed through monitoring cat numbers and assuming a direct correlation between cat numbers and wildlife deaths (Bengsen et al. 2011).

The literature generally agrees that TNR results in reduced numbers of colony cats only if immigration to the colony can be prevented (Schmidt et al. 2009; Foley et al. 2005) or if all new cats to the colony are neutered and kittens and sociable adult cats rehomed (Mendes-de-Almeida et al. 2011; Levy et al. 2003). One case study undertaken at the University of KwaZulu-Natal in South Africa, for example, showed a reduction in cat numbers when the neutering rate was 90 percent and there were no new cats joining the colonies (Jones & Downs 2011). The reduction in numbers occurs over time through natural attrition.

Colony numbers can, however, increase significantly through immigration (Gunther et al. 2011). Natoli et al. (2006), monitoring colonies in Rome, reported an immigration rate of 21 percent. Prevention of immigration is not a normal characteristic of TNR.

If one considers the reduction of feral cat numbers alone, TNR seems to be no more effective than euthanasia in closed populations and less effective in open populations (Denny & Dickman 2010; Hill 2006; Andersen et al. 2004). The cost of running a TNR program is also greater than other control methods (Jongman & Karlen 2006; Nutter et al. 2004b) and I argue that it may not be the best use of limited resources. Since it has been established that neutering does not stop hunting behaviours (Jongman & Karlen 2006) or necessarily reduce home range, increases longevity, and cats move more frequently between sites, TNR does not have any ecological benefit (Guttilla & Stapp 2010). Conservationists wish to reduce cat numbers at a landscape level whereas TNR tends to consider numbers in discrete colonies. Therefore, it is not surprising that conservationists, ecologists and TNR-proponents do not agree. TNR-proponents desire the improved welfare of feral cats but as discussed above, TNR does not necessarily result in the improvement of the welfare of feral cats.

Another important consideration is the acceptability of control programs to communities. As previously stated, terminology has a profound influence on people's opinion of control methods (Wilken 2012). Farnsworth et al. (2011) found that lethal control methods were more acceptable for feral cats than for strays. When knowledge of vulnerable wildlife is high (as in New Zealand) TNR is not supported by the majority because there is no immediate benefit to wildlife (Farnsworth et al. 2011).

Conclusion

There is no doubt that free-roaming cats have profound impacts on ecosystems through predation, spread of disease and competition, and cause public nuisance and health concerns. Also, feral cats in general suffer from poor welfare. When considering the control of these cats however, the matter becomes complex and emotive. Ethical concerns about our responsibility towards cats and towards wildlife are raised. Many animal rights and animal welfare organisations promote TNR as the best solution for cats as the alternatives involving lethal methods are viewed as unacceptable (Petersen 2012).

In urban areas where there are many willing volunteers, constant management of feral cat colonies is possible and TNR may be an effective management tool. New cats to a colony can be trapped and neutered, home ranges tend to be smaller and there are fewer vulnerable native wildlife species sharing the environment. Also, in urban settings, the attitude to lethal cat-control methods is often not positive; nevertheless, such TNR programs must include a strong volunteer base to ensure constant feeding and monitoring of the colony, high neutering rates, rehoming of kittens and sociable adult cats, and euthanasia of sick cats. In rural areas without these characteristics, TNR is neither feasible nor the best option to control feral cats.

Australia does not want to lose any more animal species so this difficult issue must be faced immediately and effectively. All relevant stakeholders must meet, listen and talk, and scientific evidence must be seen as an essential component of any discussion. It is only through such discussions that Australia can truly tackle the feral cat problem. In addition, the problem of feral cats should be tackled in other ways including the promotion of responsible cat ownership.

Works cited

Andersen MC (2007). The roles of risk assessment in the control of invasive vertebrates. *Wildlife Research*, 35: 242–48.

Andersen MC, Martin BJ & Roemer GW (2004). Use of matrix models to estimate the efficacy of euthanasia versus trap-neuter-return for management of free-roaming cats. *Journal of the American Veterinary Medical Association,* 225(12): 1871–76.

Bengsen A, Butler J & Masters P (2011). Estimating and indexing feral cat population abundances using camera traps. *Wildlife Research,* 38: 732–39.

Campbell KJ, Harper G, Algar D, Hanson CC, Keitt BS & Robinson S (2011). Review of feral cat eradications on islands. In Veitch CR, Clout MN & Towns DR (eds), *Island invasions: eradication and management* (pp37–46). Gland Switzerland: IUCN.

Coutts-Smith AJ, Mahon PS, Letnic M & Downey PO (2007). *The threat by pest animals to biodiversity in New South Wales.* Canberra: Invasive Animals Cooperative Research Centre.

Dauphine N & Cooper RJ (2009). Impacts of free-ranging domestic cats (*Felis catus*) on birds in the United States: a review of recent research with conservation and management recommendations. *Proceedings of the fourth international partners in flight conference: tundra to tropics,* 205–19.

Denny EA (2005). Ecology of free-living cats exploiting waste disposal sites: diet, morphometrics, population dynamics and population genetics. PhD thesis. University of Sydney.

Denny EA & Dickman CR (2010). *Review of cat ecology and management strategies in Australia.* Canberra: Invasive Animals Cooperative Research Centre.

Dickman CR (2007). The complex pest: interaction webs between pests and native species. In Lunney D, Eby P, Hutchings P & Burgin S (eds), *Pest or guest: the zoology of overabundance* (pp208–15). Mosman: Royal Zoological Society of New South Wales.

Farnsworth MJ, Campbell J & Adams NJ (2011). What's in a name? Perceptions of stray and feral cat welfare and control in Aotearoa, New Zealand. *Journal of Applied Animal Welfare Science,* 14(1): 59–74.

Finkler H, Hatna E & Terkel J (2011a). The influence of neighbourhood socio-demographic factors on densities of free-roaming cat populations in an urban ecosystem in Israel. *Wildlife Research,* 38: 235–43.

Finkler H, Hatna E & Terkel J (2011b). The impact of anthropogenic factors on the behaviour, reproduction, management and welfare of urban, free-roaming cat populations. *Anthrozoös,* 24(1): 31–49.

Finkler H & Terkel J (2012). Free-roaming cat overpopulation in Tel Aviv, Israel. *Preventive Veterinary Medicine,* 104(1–2): 125–35.

Foley P, Foley JE, Levy JK & Paik T (2005). Analysis of the impact of trap-neuter-return programs on populations of feral cats. *Journal of the American Veterinary Medical Association,* 227: 1775–81.

Grayson J & Calver M (2004). Regulation of domestic cat ownership to protect urban wildlife: a justification based on the precautionary principle. In D

Lunney & S Burgin (eds), *Urban wildlife: more than meets the eye* (pp169–78). Mosman: Royal Zoological Society of New South Wales.

Grayson J, Calver M & Lymbery A (2007). Species richness and community composition of passerine birds in suburban Perth: is predation by pet cats the most important factor? In D Lunney, P Eby, P Hutchings & S Burgin (eds). *Pest or guest: the zoology of overabundance* (pp195–207). Mosman: Zoological Society of New South Wales.

Gunther I, Finkler H & Terkel J (2011). Demographic differences between urban feeding groups of neutered and sexually intact free-roaming cats following a trap-neuter-return procedure. *Journal of the American Veterinary Medical Association,* 238(9): 1134–40.

Guttilla DA & Stapp P (2010). Effects of sterilization on movements of feral cats at a wildland–urban interface. *Journal of Mammalogy,* 91(2): 482–89.

Hill PM (2006). Population dynamics and management of free-roaming cats. Master's thesis. Texas A&M University.

Jessup D (2004). The welfare of feral cats and wildlife. *Journal of the American Veterinary Medical Association,* 225: 1377–83.

Jones AL & Downs CT (2011). Managing feral cats on a university's campuses: is sterilization having an effect? *Journal of Applied Animal Welfare Science,* 14: 304–20.

Jongman EC & Karlen GA (2006). Trap, neuter and release programs for cats: a literature review on an alternative control method of feral cats in defined urban areas. *Urban Animal Management Conference Proceedings:* 81–84.

Lepczyk CA, Dauphine N, Bird DM, Conant S, Cooper RJ, Duffy DC, Hatley PJ, Marra PP, Stone E & Temple SA (2010). What conservation biologists can do to counter trap-neuter-return: response to Longcore et al. *Conservation Biology,* 24(2): 627–29.

Levy JK & Crawford PC (2004). Humane strategies for controlling feral cat populations. *Journal of the American Veterinary Medical Association,* 225: 1354–60.

Levy JK, Gale DW & Gale LA (2003). Evaluation of the effect of a long-term trap-neuter-return and adoption program on a free-roaming cat population. *Journal of the American Veterinary Medical Association,* 222(1): 42–46.

Levy JK, Scott M, Lachtara JL & Crawford PC (2006). Seroprevalence of feline leukemia virus and feline immunodeficiency virus infection among cats in North America and risk factors for seropositivity. *Journal of the American Veterinary Medical Association,* 228(3): 371–76.

Liberg O, Sandell M, Pontier D & Natoli E (2000). Density, spatial organisation and reproductive tactics in the domestic cat and other felids. In Turner DC & Bateson P (eds), *The domestic cat: the biology of its behaviour* (pp119–47). Cambridge, UK: Cambridge University Press.

Little S (2011). A review of feline leukemia virus and feline immunodeficiency virus seroprevalence in cats in Canada. *Veterinary Immunology and Immunopathology,* 143: 243–45.

Longcore T, Rich C & Sullivan LM (2009). Critical assessment of claims regarding management of feral cats by trap-neuter-return. *Conservation Biology,* 23(4): 887–94.

Looney AL, Bohling MW, Bushby PA , Howe LM, Griffin B, Levy JK, Eddlestone SM, Weedon JR, Appel LD, Rigdon-Brestle K, Ferguson NJ, Sweeney DJ, Tyson KA, Voors AH, White SC, Wilford CL, Farrell KA, Jefferson EP, Moyer MR, Newbury SP, Saxton MA & Scarlett JM (2008). The Sssociation of Shelter Veterinarians veterinary medical care guidelines for spay-neuter programs. *Journal of the American Veterinary Medical Association,* 233(1): 74–86.

Lord LK (2008). Attitudes and perceptions of free-roaming cats among individuals in Ohio. *Journal of the American Veterinary Medical Association,* 232(8): 1159–67.

Lowe S, Browne M, Boudjelas S & De Poorter M (2000). *100 of the world's worst invasive alien species: a selection from the global invasive species database.* Auckland, New Zealand: Invasive Species Specialist Group, International Union for Conservation of Nature.

Loyd KA & DeVore JL (2010). An analysis of feral cat management options using a decision analysis network. *Ecology and Society,* 15(4): 10. [Online] Available: www.ecologyandsociety.org/vol15/iss4/art10/ [Accessed 11 September 2012].

Loyd KA & Hernandez SM (2012). Public perceptions of domestic cats and preferences for feral cat management in the southeastern United States. *Anthrozoös,* 25(3):337–51.

Loyd KA & Miller CA (2010a). Factors related to preferences for trap-neuter-release management of feral cats among Illinois homeowners. *Journal of Wildlife Management,* 74(1): 160–65.

Loyd KA & Miller CA (2010b). Influence of demographics, experience and value orientations on preferences for lethal management of feral cats. *Human Dimensions of Wildlife: An International Journal,* 15(4): 262–73.

Manfredo M, Teel T & Bright A (2003). Why are public values towards wildlife changing? *Human Dimensions of Wildlife: An International Journal,* 8(4): 287–306.

Mendes-de-Almeida F, Labarthe N , Guerrero J, Faria MCF, Branco AS, Pereira CD, Barreira JD & Pereira MJS (2007). Follow-up of the health conditions of an urban colony of free-roaming cats (*Felis catus* Linnaeus 1758) in the city of Rio de Janeiro, Brazil. *Veterinary Parasitology,* 147: 9–15.

Mendes-de-Almeida F, Remy G & Gershony LC et al. (2011). Reduction of feral cat (*Felis catus* Linnaeus 1758) colony size following hysterectomy of adult female cats. *Journal of Feline Medicine and Surgery,* 13(6): 438–40.

Moodie E (1995). *The potential for biological control of feral cats in Australia.* Canberra: Australian Nature Conservation Agency.

National Consultative Committee on Animal Welfare (2008). The welfare of cats, the NCCAW position statement. [Online] Available: www.daff.gov.au/animal-plant-health/welfare/nccaw/guidelines/pets/cats [Accessed 1 April 2010].

Natoli E, Maragliano L , Cariola G, Faini A, Bonanni R, Cafazzo S & Fantini C (2006). Management of feral cats in the urban environment of Rome (Italy). *Preventative Veterinary Medicine*, 77: 180–85.

Nutter FB, Levine JF & Stoskopf MK (2004a). Reproductive capacity of free-roaming domestic cats and kitten survival rate. *Journal of the American Veterinary Medical Association,* 225: 1399–402.

Nutter FB, Stoskopf KK & Levine JF (2004b). Time and financial costs of programs for live trapping feral cats. *Journal of the American Veterinary Medical Association*, 225: 1403–05.

Petersen N (2012). Talking TNR: Promoting a better approach to feral cats to your local officials. *Animal Sheltering,* July/August: 41–46.

Robertson SA (2008). A review of feral cat control. *Journal of Feline Medicine and Surgery,* 10: 366–75.

Schmidt PM, Lopez RR & Collier BA (2007). Survival, fecundity, and movements of free-roaming cats. *Journal of Wildlife Management,* 71(3): 915–19.

Schmidt PM, Swannack TM, Lopez RR & Slater MR (2009). Evaluation of euthanasia and trap-neuter-return (TNR) programs in managing free-roaming populations. *Wildlife Research,* 36: 117–25.

Scott KC, Levy JK & Crawford C (2002). Characteristics of free-roaming cats evaluated in a trap-neuter-return program. *Journal of the American Veterinary Medical Association,* 221: 1136–38.

Slater MR (2002). *Community approaches to feral cats: problems, alternatives & recommendations.* Washington: Humane Society Press.

Slater MR (2005). The welfare of feral cats. In I Rochlitz (ed.), *The welfare of cats* (pp141–76). London: Springer.

Spada E, Proverbio D, della Pepa A, Perego R, Baggiani L, De Giorgi GB, Domenichini G, Ferro E & Cremonesi F (2012). Seroprevalence of feline immunodeficiency virus, feline leukaemia virus and *Toxoplasma gondii* in stray cat colonies in northern Italy and correlation with clinical laboratory data. *Journal of Feline Medicine and Surgery*, 14(8): 369–77.

Teel TL, Manfredo MJ & Stinchfield HM (2007). The need and theoretical basis for exploring wildlife value orientations cross-culturally. *Human Dimensions of Wildlife: An International Journal,* 12(5): 297–305.

Tennent J, Downs CT & Bodasing M (2009). Management recommendations for feral cats (*Felis catus*) populations within an urban conservancy in

KwaZulu-Natal, South Africa. *South African Journal of Wildlife Research,* 39(2): 137–42.

van Heezik Y (2010). Pussyfooting around the issue of cat predation in urban areas. *Oryx,* 44(2): 153–54.

Wallace JL & Levy JK (2006). Population characteristics of feral cats admitted to seven trap-neuter-return programs in the United States. *Journal of Feline Medicine and Surgery,* 8(4): 279–84.

West P (2008). *Assessing invasive animals in Australia 2008.* Braddon, ACT: Invasive Animals Cooperative Research Centre, National Land & Water Resources Audit.

Wilken RL (2012). Feral cat management: perceptions and preferences (a case study). Master's thesis, paper 4181. San Jose State University.

Winograd NJ (2007). *Redemption: the myth of pet overpopulation and the no kill revolution in America.* The United States of America: Almaden Books.

Witmer G, Constantin B & Boyd F (2005). Feral and introduced carnivores: issues and challenges. *Wildlife damage management conferences – proceedings.* Paper 86.

11

Animal farming in Australia: consumer awareness, concern and action

Sally Healy

Humans have come a long way from our hunter-gatherer origins. Not only do we eat more animal protein than ever before but, due to rapid advances in technology, animal agriculture has evolved into a highly efficient production system that has the potential to support the world's growing population. Australians are among the biggest per capita meat-eaters in the world (FAO 2009). Between 2003 and 2004, the Australian agriculture industry slaughtered 24.1 million cattle, 2.55 million pigs, 419 million poultry (of different species), and 94 million sheep (Department of Agriculture, Fisheries and Forestry 2005). Intensive animal production, also known as factory farming, is highly efficient, automated and mechanised, ensuring an economic advantage over the traditional family farm (Garner 2004). It is also associated with negative animal welfare.

With a growing shift towards intensive farming systems, there is substantial consumer concern for the ethical dimension of modern farming systems, in particular the animal welfare implications of modern farming methods. Pork, chicken and eggs are predominately sourced from intensive production systems in Australia which in turn have created a niche market segment based on alternative, more ethical, production methods, such as free-range, organic and bred free-range. Literature in the area of food production has identified several key determinants of welfare-friendly consumption behaviours, including socio-demographics, beliefs, social norms and product perceptions.

185

One factor that has received limited attention, however, is consumer knowledge of animal welfare conditions. Furthermore, the relationship between knowledge and consumption choices is relatively unexplored despite the pertinence of this issue for a range of stakeholders in animal farming.

This chapter is concerned with the results of an empirical study that was designed to investigate the relationship between Australian consumers' knowledge of modern farming practices and their consumption of animal-based foods. Framed within the theory of planned behaviour, the study examined consumer knowledge, perceptions, and preferences for eggs, pork, and chicken products. The purpose of the study was to analyse both self-reported and actual levels of welfare knowledge and relate these to reported consumption behaviours. The findings suggest that Australian consumers are concerned with the welfare implications of intensive farming systems but do not have adequate information on farm-animal welfare and the meanings behind value-based food labels.

The cost of intensive systems

As societies become wealthier and more populated, there is increased pressure on producers and suppliers to provide adequate sustenance to the general population. Intensive animal-agricultural systems make up the bulk of animal farming in the developed world and are growing at a rapid rate in the developing world. Intensive animal-agriculture systems have gained a negative image among consumers due to their belief that a certain level of efficiency is detrimental to the animals involved, particularly features such high stocking densities, close confinement, body modifications, and limited genetic diversity that are characteristic of such designs (Loughnan 2012; Williams 2008).

Despite concerns surrounding animal agriculture, research in this area indicates that consumers do not generally have adequate knowledge of the welfare implications of such systems and the welfare-friendly alternative products available (Vanhonacker et al. 2010; McEachern & Warnaby 2008). This may reflect either a lack of accurate information about agricultural practices or the suppression of a psychological and cultural connection between concerns for animal welfare and behaviour. The sociology of denial has contributed to our understanding of the psychological and social mechanisms that allow consumers to remain unaware or unaffected by information on animal suffering (Wicks 2011). The inclination for consumers to maintain a barrier between themselves and unpleasant information on farming systems may allow them to continue to make choices based on other product attributes such as price, quality and accessibility. Unsatisfactory welfare conditions for farm animals may be rendered acceptable to the public through mechanisms of denial and the selected avoidance of unpleasant information (Williams 2008).

Concern for animal welfare

Despite the ubiquity of intensive animal-agrarian systems, research suggests animal welfare is a concern for many consumers. In Franklin's (2007) survey of over 2000 Australians, 42 percent reported feeling anxiety about the meat industry and 27 percent reported being less likely to eat meat than they were a few years ago. Respondents reported discomfort with the modern factory-style farming, with over half of the 2000 respondents agreeing that such systems are 'unnatural' and 'cruel'. Southwell et al. (2006) also found that consumers felt concerned about the treatment and transport of livestock. In Franklin's study, respondents reported favourable attitudes towards the concept of meat-eating provided the animals were treated humanely throughout the production line.

The emergence of consumer concern for the ethical dimension of modern farming is relevant for several stakeholder groups. Firstly, consumers have the capacity to impact the supply chain by demanding changes to production standards. Secondly, industry bodies and producers must remain aware of and responsive to consumer concerns whilst maintaining highly efficient, innovative production systems and complying with regulations (Verbeke 2000). Finally, governments must respond to consumer pressure, provide support to industry and producers, and provide sufficient legislation to protect the interests of consumers, farmers, and livestock animals. The dramatic increase in research on consumer perceptions, attitudes, and preferences over the last several years is indicative of the far-reaching implications of changing perceptions.

The dynamic between producers and consumers is reflective of an inherent conflict of interest between the two groups. Producers must survive under intense pressures from consumers for affordable, safe, and high-quality animal-based foods. Advances in science and technology have enabled an efficient system of food production; however, this is often at the expense of animal welfare. Producers are therefore conflicted between running a highly efficient system and meeting consumer demands for animal-friendly products. In light of increased consumer concern and awareness of animal welfare issues inherent in modern farming systems, farmers and producers attempt to encourage a positive perception by consumers. They do this through communicating with the media and directly with consumers through packaging, promotion, and accreditation (Dentoni et al. 2011).

The increasing consumer preference for ethically sourced goods prompts some brands to create specialty products, thus allowing them to charge a higher price in exchange for satisfying the consumer's desire for supporting superior animal welfare or environmental sustainability values. A common strategy for existing brands is to extend into new product categories (Aaker & Keller 1990). An example of a brand extension is RSPCA UK's Freedom Food label, which encourages stakeholders along the farming supply chain, including producers and retailers, to participate provided they meet certain welfare standards. Value-based labels on animal-based foods are an important way for brands to communicate with the consumer on the ethical viability of the production and quality of the product (Barham 2002). In their study of value-based labels, McEachern and Warnaby (2008) found that consumers had little understanding of the standards represented by the RSPCA UK's Freedom Food label, but viewed the products in a positive way – reflecting the benefit of the product's association with the organisation's favourable reputation.

The rise of ethical consumerism

Ethical consumerism is a rapidly developing social movement in which consumption choices are influenced by concerns on the moral, social, and/or ethical implications of production (Carrigan et al. 2004). The ethical consumer considers the impacts of their consumption behaviours on society, animals, and the environment: many people now communicate their ethical ideology through their consumption behaviours (Carrier 2007; Uusitalo & Oksanen 2004). One of the primary goals of ethical-consumerism research is to identify the factors that motivate concerned citizens to express activist behaviours in the form of modified purchasing behaviours.

One stream of ethical consumerism is concern for the animal welfare implications of food production. Animal welfare is a major issue that affects the entire agriculture industry, including farmers, retailers, consumers, and animals. The literature indicates an increase in consumer concern over the last several years which has given rise to more vigorous research efforts to investigate the way consumers act on concerns and the relationship between legislation, attitudes, intent, and behaviour (Bennett & Blaney 2003). The rise in consumer concern for the welfare of animals in the farming industry can be attributed to the growing shift to intensive farming methods as well as numerous public controversies such as live-animal export and documented cruelty to animals in abattoirs – causing what Parbery and Wilkinson (2012, 1) term a change in the 'social authorising environment'. Public attention to cruelty or mistreatment in agriculture opens up the potential for citizens to become activists by choosing alternative foods or products that support animal welfare–friendly modes of production.

Much of the research in the field of human relationships with animals has drawn upon empirical and theoretical studies to examine the most important factors in predicting attitudes towards animals. The consensus is that the relationship between attitudes towards animals and behaviour is complex and dependent on the personality, psychometric and demographic characteristics of the individual, as well as the species of animal (Frewer et al. 2005). Beyond the level of individual characteristics, it is likely that external factors, for example education, experience, knowledge and social climate, have the potential to influence attitudes and in turn behaviour (Serpell 2004).

Theoretical considerations

Research in the field of consumer behaviour is largely influenced by theoretical frameworks in which attitudes and beliefs play a significant role in determining intention and subsequent behaviour (Robinson & Smith 2002). Ajzen's (1991) theory of planned behaviour (TPB) is an extension of Fishbein and Ajzen's (1975) theory of reason action (TRA), which suggests that beliefs influence both attitudes and subjective norms, which in turn determine whether the consumer intends to purchase a particular product. The TPB incorporates the consumer's perceived behavioural control – a measure that assesses the degree to which the consumer feels the purchase decision is under their volitional control.

The study reported in this chapter was framed around the TPB, not only because the theory has demonstrated predictive capacity in behavioural research in a diverse range of disciplines (Armitage & Conner 1999; Sheppard et al. 1988), but also because previous research on the relationship between beliefs – which are a function of knowledge – and actual behaviour is relatively underdeveloped. The three determinants of intention – attitudes, subjective norms and perceived behavioural control – were a focus of the study, with a particular emphasis on consumers' level of claimed and actual knowledge, which underpins beliefs on the treatment of farming animals and their welfare needs.

Attitudes are a fundamental component of the TPB; they indicate the consumer's evaluation of certain product attributes or behaviours (Vermeir & Verbeke 2008). It is important to note, however, that positive attitudes towards a certain practice – in this case ethical methods of animal agriculture – do not imply those attitudes will translate into direct behaviour (Vermeir & Verbeke 2006). Indeed, significant research effort is devoted to understanding the intention–behaviour gap observed in human behaviour studies (Vitell 2003). Subjective norms refer to the consumer's perception of the behaviours of those around them, and the degree to which perceived opinions influence purchasing decisions (Vermeir & Verbeke 2008; Cialdini & Goldstein 2004).

Consumer awareness of animal welfare conditions is an overlooked area within ethical consumerism research. Although animal products are readily available, there is evidence of a significant disconnect between the consumer and the choice of products available (Vanhonacker et al. 2010). Based on research to date, the typical consumer has a limited understanding of the moral, environmental and social implic-ations of modern systems that produce animal-based food products. Knowledge is recognised as an important component of the consumer decision-making process, particularly the mediating impact of know-ledge on attitudes towards the treatment of farm animals and subse-quent purchasing decisions. The stages of information processing in-clude the consumer's exposure to information – usually from a combin-ation of sources such as the general media, animal-protection groups, and product advertising and labelling – through to comprehension, persuasion, and retention, which often involves a reappraisal of current behaviour (Verbeke 2000). While the rise in concern for animal welfare among consumers and the factors that influence this concern – includ-ing knowledge of animal welfare – are well documented, less attention has been paid to the relationship between knowledge and behaviour (McEachern & Warnaby 2008). Furthermore, the existing research is focused on consumers' knowledge of value-based labels rather than implicit knowledge of welfare needs of farm animals (Lawson 2002; Ratchford 2001).

The majority of animal-based foods are sourced from intensive systems which raises ethical concerns for some consumers. Although there is evidence for a lack of consumer confidence in farm-animal welfare and a desire for greater transparency in the food-production industry, little is known about the role of awareness, experience, and so-cial climate in shaping behaviour related to concern for animal welfare. Guided by the theoretical considerations dominant in the literature, a survey was devised to examine the prevalence of concern for farm-animal welfare among Australian consumers as well as the breadth of knowledge that consumers have on animal farming.

Methods

Surveying Australian consumers

An online survey was developed to collect data on the relationship between knowledge, attitudes, perceptions, and consumption of animal-based foods. The survey was open for two months in early 2012 to anyone over the age of 16 and was distributed through a snowball sampling method. The survey was advertised to staff and students at two Australian universities via email lists, as well as the social networking site Facebook. Participants were invited to forward the survey to acquaintances in their social and professional networks. Online surveying was selected as the mode of participant recruitment because it is cost and time efficient and allows the researcher to gather adequate data on the topic from participants (Evans & Mathur 2005).

A total of 840 people completed the survey and results were analysed using Statistical Package for Social Sciences (SPSS 2004). The survey consisted of both open and closed questions to investigate five broad categories relevant to the survey objectives and reflective of the theory of planned behaviour: current consumption behaviour, knowledge of animal-agriculture practices, attitudes and preferences, subjective norms and perceived behavioural control, and socio-demographic characteristics. The majority of respondents were female (80 percent), aged between 19 and 34 (61 percent), and had completed tertiary education (56 percent).

Consumption behaviour

Current consumption behaviour that was examined included frequency of meat and dairy intake, type of diet followed (meat-eater, vegetarian, vegan), and reason for avoiding meat if applicable. Participants were also asked to indicate product labels they look for when buying eggs and meat; for example, free-range, organic, or biodynamic.

Knowledge

A claimed knowledge score was determined by asking participants to assess their level of understanding of modern animal-based food production techniques on a five-point unipolar scale ranging from 'no understanding' to 'extensive understanding'. Actual knowledge was gauged using three sections of questions. The first section comprised a list of statements about Australian animal farming from which participants were asked to indicate whether the statement was true, false or if they were uncertain. Example questions from this section included: 'Beak trimming in laying hens is permitted without pain relief'; 'hormones are added to chicken feed'; and 'pigs raised intensively spend the majority of their lives indoors'.

The next section had contained a list of five practices from which participants were asked to indicate whether or not they are permitted on Australian livestock. Statements included: 'piglets – tails clipped'; 'hens – forced moulting'. The final section followed the same format as the first; however the statements reflected natural behaviours and welfare requirements for pigs and chickens raised for farming. Examples are 'Pigs are largely inactive and spend most of their time sleeping' and 'the natural lifespan of a chicken is 2–3 years'. Actual knowledge scores were calculated for each participant based on the number of correct responses with a possible score between zero and 22.

Attitudes and preferences

Attitudes were measured using 'profit' statements from the pest, pet, profit (PPP) scale as developed by Taylor and Signal (2009a) as well as five additional statements that referenced the viability of factory farming and the moral worth of livestock animals. Participants were asked to indicate their agreement with each statement on a five-point scale ranging from strongly agree to strongly disagree. To understand the significance of animal welfare on product selection, participants were asked to rank the five most important attributes for eggs, pork and chicken. Attributes included: Australian in origin, brand, concern for environment, price, and free-range.

Subjective norms and perceived behavioural control

Normative beliefs were assessed using two statements:

1. Most people who are important to me think that buying animal-based foods that promote acceptable treatment of animals is something I should do.
2. Most people make an effort to buy animal-based foods that promote the acceptable use of animals.

The responses were collected using a Likert scale from 'strongly agree' to 'strongly disagree'. To assess perceived behavioural control, participants completed the same scale for the question 'whether or not I consume animal-based foods that are produced using methods that promote acceptable treatment of animals is completely up to me'. Respondents also indicated how easy it is for them to find animal-friendly foods based on labels and their concern for the treatment of farm animals.

Socio-demographics

Finally, participants provided demographic information such as age, gender, postcode, experience with pets and livestock, ethnicity, religion, education, and income. These variables were included in the survey so that the relationships between socio-demographics and concern for animal welfare could be investigated.

Understanding Australian consumers

Responses from the 840 participants suggest that animal welfare is important to consumers. Just over half of respondents (57 percent) claimed to be either quite concerned or extremely concerned about the treatment of farm animals in Australia. Over 84 percent of participants agreed with the statement 'It is cruel to keep birds in cages to mass produce eggs', and almost 70 percent agreed with the statement 'Modern methods of "factory farming" to produce eggs, milk and meat are cruel'. It is likely that the emotive wording used in the latter two statements prompted survey participants to respond in a certain way. The question 'how concerned are you regarding the treatment of farm animals in Australia?' possibly invited respondents to think of farming based on personal experiences rather than the negative connotations of 'cages' and 'factory farming' as used in alternative questions. Overall, however, these findings are consistent with other studies that explore consumer concern for animal welfare. Harper and Henson's (2001) study of European attitudes to farm-animal welfare, for example, reported high levels of concern about farming standards due to the animal welfare implications of intensive systems and a consensus of 'high level of concern' has been met across the literature (Toma et al. 2012; Martelli 2009).

There was a significant (p < 0.05) relationship between concern for farm animal treatment and consumption, with vegetarians and vegans reporting higher concern than meat-eaters. Religion was the only socio-demographic variable that was strongly related with concern for the treatment of farm animals. This is in contrast to findings which have suggested that females, those with lower incomes and education levels, and younger and middle-aged people have a higher concern for animal welfare (Kendall et al. 2006). The findings from this study support the claim that no consensus can be met in terms of the capacity for socio-demographic variables to predict attitudes.

Knowledge

Participants who claimed to have high knowledge generally earned higher knowledge scores; vegans and vegetarians reported higher self-rated knowledge of animal farming than meat-eaters (p<0.05) as determined by a chi-squared test. Actual knowledge scores were calculated for each participant, with a possible score between zero and 22. There was a significant difference between consumer groups (meat-eaters, vegetarians, and vegans) as determined by the Welch *t* test. A Tukey post-hoc test revealed significant difference in knowledge scores between all three consumer groups, with higher knowledge scores associated with lower levels of consumption of animal-based foods.

Gender and religion were correlated with consumer behaviour, however ethnic background, income, education and age were not. The relationship between gender and vegetarianism (and veganism) is to be expected, as it is well established in the literature that females are more likely to abstain from meat consumption (Beardsworth & Bryman 2004; Kubberød et al. 2002). Several studies have shown a negative correlation between religiosity and concern for animal welfare, which is likely to influence diet (Heleski et al. 2006). Although religion has been associated with food choices, it generally explains little of the overall variance in consumption choices and ethical concerns about animal welfare (Taylor & Signal 2009b; Honkanen et al. 2006).

Actual knowledge scores correlated with age and with whether the participant had lived on a livestock farm. Those who had lived on a livestock farm, however, were less likely to follow a vegetarian or vegan diet than those who had not. This finding differs from the study by Parbery and Wilkinson (2012), which determined that both knowledge and experience with farming were associated with higher levels of critical activism, which was defined as protesting or altering shopping habits as an expression of discontent with current practices. That those with exposure to farming had higher knowledge yet were less likely to abstain from animal-based foods suggests that the impact of such experience could be mediated through processes of socialisation and desensitisation, and the acquisition of particular values relevant to a farming lifestyle (Willits & Luloff 1995).

Consumption behaviour

Forty-five percent of participants were not restrictive in their food choices. Thirty-six percent claimed to avoid some types of meat but did not identify themselves as vegetarian, while six percent considered themselves a vegetarian but consumed some types of meat. Eight percent identified as strict vegetarian (no meat) and five percent were vegan. Of those participants who identified as vegetarian or vegan, the most commonly cited reason for doing so was concern with the way animals are treated (82 percent), animal rights (74 percent), concern for the environment (60 percent), health (49 percent) and religion/upbringing (<5 percent).

Product preferences and attitudes

The apparent interest in animal welfare was also reflected in participants' responses to product preferences. When participants were asked to select labels they looked for when purchasing eggs, 'free-range' was the most desired label (72 percent), followed by 'organic' (27 percent). Similarly for meat, 36 percent of participants selected free-range as their preferred option. Product selection is largely influenced by evaluations of product attributes and their relative importance to the consumer, based on individual values, attitudes, and needs (Fishbein 1967). Information has the potential to change beliefs about, and evaluations of, a product. The strong preference of respondents to prefer free-range egg and meat products is likely due to the increased media coverage and public awareness of the term 'free-range' and its ethical implications in recent years (Tonsor & Olynk 2011). It is interesting that only 57 percent of respondents reported a concern for the treatment of farm animals, yet 72 percent claim to prefer free-range eggs. It could be that consumers do not feel as concerned about livestock welfare when there are high welfare products available. The discrepancy could also be explained by the intention–behaviour gap, as often noted in ethical consumerism research (Bray et al. 2011).

Perceived behavioural control and subjective norms have been explored in previous studies on ethical consumerism and often highlight the importance of social climate on purchasing decisions. As Fishbein and Ajzen (1975) state, consumer attitudes are more strongly associated with behavioural intent when subjective norms are supportive of those beliefs. It is noteworthy that 63 percent of participants in this study do not think that most people make an effort to buy high welfare products. Sixty-nine percent agree that it is completely up to them whether they buy animal welfare–friendly products. However, almost 58 percent claimed that it is difficult to identify animal-based foods that promote the acceptable treatment of animals. This is slightly higher than was found for European Union citizens, of which 51 percent claimed to have difficulty understanding how animal-friendly products are, based on their label (Martelli 2009). Consumers' lack of understanding of the meaning behind food labels affects ethical purchasing decisions because the label is an important medium for communicating information about the product (Verbeke 2009; Howard & Allen 2006). These results indicate that consumers perceive the choice of ethical purchasing decisions to be mostly theirs, but feel that others do not make the same effort or perhaps do not care as much.

Bridging the gap between concern and behaviour

Consumer trust in farmers and producers is relevant to understanding ethical consumer behaviour. Some studies indicate a growing distrust among consumers towards the credibility of ethical labelling schemes of animal products (Frewer et al. 2005). In this study 57 percent of participants disagreed with the statement 'The meat production and processing industries can be trusted to ensure the safety of the meat product'. This is significantly higher than the 31 percent of respondents who responded to the same item in Franklin's 2007 study. Trust is a recurring theme in discussions on farming and animal welfare. A study by Nocella et al. (2010) demonstrated an increase in consumer willingness to pay for improved animal welfare when trust in stakeholders within the supply chain was established. An important message taken from the findings of this study, and others of similar nature, is that trust in the supply chain is highly important to consumer confidence. As well as animal welfare considerations, consumers desire access to accurate information on the attributes of food items, including genetic modification, chemical usage and sustainability (Hoogland et al. 2005).

Willingness to pay for improved animal welfare was related to income and gender, which differs from findings by Zhao and Wu (2012), that willingness to pay was related to age, education, and income, but not gender. Concern with animal welfare and willingness to pay varies widely between socio-demographic characteristics and there is no clear market segment that would benefit most from increased transparency of the food chain.

Knowledge, subjective norms, and perceived behavioural control all have an effect on ethical purchasing behaviours. The results of this study show that higher levels of farming knowledge as it relates to animal welfare are correlated with decreased consumption of animal-based foods. This finding may be indicative of a diminishing level of acceptability of intensive farming practices. It is in the best interests of the relevant stakeholders to remain aware of consumer attitudes, preferences and concerns, and communicate effectively to consumers through available outlets – including brand extension, accreditation, and labelling – in order to restore trust and remain competitive in the market. These findings reinforce the need for agricultural groups to respond to concerns related to animal welfare from consumers and other stakeholders, and to allow sufficient transparency in welfare considerations and label and accreditation schemes in order to improve the public's trust in Australian farming.

Transparency is an important concept for understanding ethical consumer behaviour. In the ideal state, people have access to adequate information about food production and they can therefore make choices based on matching their ethical values with the values represented by competing companies and producers (Hoogland et al. 2005). There must be a certain level of transparency in the farming system in order for consumers to make an informed decision on the products available (Vanhonacker et al. 2010; Hoogland et al. 2005). The confusion respondents of this study reported about the animal welfare information of product labels indicates the need for clear, accurate information at the point of purchase. This study has shown that knowledge of farming conditions and understanding of product labels are likely to influence consumers' willingness to purchase animal welfare-friendly food products.

It is important, however, to recognise that concern does not directly translate to ethical purchase behaviour. The gap between intention to buy and actual purchase is well documented, for example, the literature on ethical consumerism details how individuals may feel compelled to answer in a way that they perceive is socially desirable (Freestone & McGoldrick 2008). The tendency for participants to report behaviours they perceive as being ethical emphasises the need for studies that measure actual behaviour rather than reported behaviour or intention. Toma et al. (2010) cite two possible reasons for the gap between intention and actual behaviour: lack of information on animal welfare and perception of labelling. This explanation is consistent with the findings from this study, as those participants with more accurate knowledge of farming were more likely to consume limited animal-based foods.

Conclusion

This study has made several key findings that are relevant for the stakeholders involved in livestock production in Australia and abroad. Similar to other research efforts, the majority of respondents reported at least some concern on the animal welfare conditions for livestock. Despite this concern, many consumers found it difficult to vote with their dollar to promote animal welfare because food labels were confusing, resulting in difficulty discerning whether or not a producer promoted high welfare conditions. This finding highlights the importance of consumer information in the decision-making process and the need for transparency in the food production industry. The correlation between knowledge and decreased consumption of animal products suggests there is concern for the ethics of intensive farming. Importantly, the relationship between farming knowledge and experience with livestock indicates the importance of other social and cultural factors in influencing concern and consumer choices.

While intensive farming systems have the capacity to provide large quantities of animal-based foods to our growing population, the negative impacts on farm-animal welfare are a concern to Australian consumers. Consumer demand for alternative, more ethically sourced foods is a step in the right direction. However, as this study has shown, Australians are not as connected to food production as they would have been several decades ago. There is a growing discomfort about the treatment of livestock, yet confusion about labelling makes it difficult for consumers to make an informed choice in favour of high welfare products. Consumers require accurate information and, if possible, experience with livestock to be in the best position to make ethical purchasing decisions. With this information and understanding Australian consumers will once again become connected to the foods they eat and enjoy comfort in knowing that animal welfare is a priority for modern farming.

Works cited

Aaker DA & Keller KL (1990). Consumer evaluations of brand extensions. *Journal of Marketing*, 54(1): 27–41.

Ajzen I (1991). The theory of planned behavior. *Organizational Behavior and Human Decision Processes*, 50(2): 179–211.

Armitage CJ & Conner M (1999). The theory of planned behaviour: assessment of predictive validity 'perceived control'. *The British Journal of Social Psychology*, 38: 35–54.

Barham E (2002). Towards a theory of values-based labeling. *Agriculture and Human Values*, 19(4): 349–60.

Beardsworth A & Bryman A (2004). Meat consumption and meat avoidance among young people. An 11-year longitudinal study. *British Food Journal*, 106(4): 313–27.

Bennett RM & Blaney RJP (2003). Estimating the benefits of farm animal welfare legislation using the contingent valuation method. *Agricultural Economics*, 29(1): 85–98.

Bray J, Johns N & Kilburn D (2011). An exploratory study into the factors impeding ethical consumption. *Journal of Business Ethics*, 98(4): 597–608.

Carrier JG (2007). Ethical consumption. *Anthropology Today*, 23(4): 1–2.

Carrigan M, Szmigin I & Wright J (2004). Shopping for a better world? An interpretive study of the potential for ethical consumption within the older market. *The Journal of Consumer Marketing*, 21(6): 401–17.

Cialdini RB & Goldstein NJ (2004). Social influence: compliance and conformity. *Annual Review of Psychology*, 55(1): 591–621.

Dentoni D, Tonsor G, Calantone R & Peterson C (2011). 'Animal welfare' practices along the food chain: how does negative and positive information affect consumers? *Journal of Food Products Marketing*, 17: 279–302.

Department of Agriculture, Fisheries and Forestry (2005). *Australian agriculture and food sector stocktake*. Canberra: Commonwealth Government.

Evans J & Mathur A (2005). The value of online surveys. *Internet Research*, 15(2): 195–219.

FAO *see* Food and Agriculture Organization of the United Nations.

Fishbein M (1967). A behaviour theory approach to the relations between beliefs about an object and the attitude toward the object. In M Fishbein (ed), *Readings in attitude theory and measurement* (pp389–400). New York: John Wiley & Sons.

Fishbein M & Ajzen I (1975). *Belief, attitude, intention and behaviour: an introduction to theory and research*. Reading, MA: Addison-Wesley.

Food and Agriculture Organization of the United Nations (2009). *The state of food and agriculture: livestock in the balance*. Rome: Food and Agriculture Organization of the United Nations.

Franklin A (2007). Human–nonhuman animal relationships in Australia: an overview of results from the first national survey and follow-up case studies 2000–2004. *Society and Animals*, 15: 7–27.

Freestone OM & McGoldrick PJ (2008). Motivations of the ethical consumer. *Journal of Business Ethics*, 79(4): 445–67.

Frewer LJ, Kole A, van der Kroon SMAV & de Lauwere C (2005). Consumer attitudes towards the development of animal-friendly husbandry systems. *Journal of Agricultural and Environmental Ethics*, 18(4): 345–67.

Garner R (2004). *Animals, politics and morality*. Manchester, UK: Oxford University Press.

Harper G & Henson S (2001). *Consumer concerns about animal welfare and the impact on food choice*. Reading: Centre for Food Economics Research, Department of Agricultural and Food Economics, The University of Reading.

Heleski CR, Mertig AG & Zanella AJ (2006). Stakeholder attitudes toward farm animal welfare. *Anthrozoös*, 19(4): 290–307.

Honkanen P, Verplanken B & Olsen SO (2006). Ethical values and motives driving organic food choice. *Journal of Consumer Behaviour*, 5(5): 420–30.

Hoogland CT, de Boer J & Boersema JJ (2005). Transparency of the meat chain in the light of food culture and history. *Appetite*, 45(1): 15–23.

Howard PH & Allen P (2006). Beyond organic: consumer interest in new labelling schemes in the Central Coast of California. *International Journal of Consumer Studies*, 30(5): 439–51.

Kendall HA, Lobao LM & Sharp JS (2006). Public concern with animal well-being: place, social structural location, and individual experience. *Rural Sociology*, 71(3): 399–428.

Kubberød E, Ueland Ø, Rødbotten M, Westad F & Risvik E (2002). Gender specific preferences and attitudes towards meat. *Food Quality and Preference*, 13(5): 285–94.

Lawson R (2002). Consumer knowledge structures: background issues and introduction. *Psychology & Marketing*, 19(6): 447–56.

Loughnan D (2012). *Food shock*. Wollombi, NSW: Exisle Publishing Pty Ltd.

Martelli G (2009). Consumers' perception of farm animal welfare: an Italian and European perspective. *Italian Journal of Animal Science*, 8(1): 31–41.

McEachern MG & Warnaby G (2008). Exploring the relationship between consumer knowledge and purchase behaviour of value-based labels. *International Journal of Consumer Studies*, 32(5): 414–26.

Nocella G, Hubbard L & Scarpa R (2010). Farm animal welfare, consumer willingness to pay, and trust: results of a cross-national survey. *Applied Economic Perspectives and Policy*, 32(2): 275–97.

Parbery P & Wilkinson R (2012). *Victorians' attitudes to farming*. Melbourne, Victoria: Department of Primary Industries.

Ratchford BT (2001). The economics of consumer knowledge. *Journal of Consumer Research*, 27(4): 397–411.

Robinson R & Smith C (2002). Psychosocial and demographic variables associated with consumer intention to purchase sustainably produced foods as defined by the Midwest Food Alliance. *Journal of Nutrition Education and Behavior*, 34(6): 316–25.

Serpell JA (2004). Factors influencing human attitudes to animals and their welfare. *Animal Welfare*, 13: S145–51.

Sheppard BH, Jon H & Warshaw PR (1988). The theory of reasoned action: a meta-analysis of past research with recommendations for modifications and future research. *Journal of Consumer Research*, 15(3): 325–43.

Southwell A, Bessey A & Barker B (2006). *Attitudes towards animal welfare: a research report*. Canberra: Department of Agriculture, Fisheries and Forestry.

Statistical Package for Social Sciences (2004). SPSS for Windows, Version 12.0. Chicago, SPSS Inc.

Taylor N & Signal T (2009a). Pet, pest, profit: isolating differences in attitudes towards the treatment of animals. *Anthrozoös*, 22(2): 129–35.

Taylor N & Signal TD (2009b). Willingness to pay: Australian consumers and 'on the farm' welfare. *Journal of Applied Animal Welfare Science*, 12(4): 345–59.

Toma L, Kupiec-Teahan B, Stott AW & Revoredo-Giha C (eds) (2010). Animal welfare, information and consumer behaviour. 9th European IFSA Symposium; Vienna.

Toma L, Stott AW, Revoredo-Giha C & Kupiec-Teahan B (2012). Consumers and animal welfare. A comparison between European Union countries. *Appetite*, 58(2): 597–607.

Tonsor GT & Olynk NJ (2011). Impacts of animal well-being and welfare media on meat demand. *Journal of Agricultural Economics*, 62(1): 59–72.

Uusitalo O & Oksanen R (2004). Ethical consumerism: a view from Finland. *International Journal of Consumer Studies*, 28(3): 214–21.

Vanhonacker F, Van Poucke E, Tuyttens F & Verbeke W (2010). Citizens' views on farm animal welfare and related information provision: exploratory insights from Flanders, Belgium. *Journal of Agricultural and Environmental Ethics*, 23: 551–69.

Verbeke W (2000). Influences on the consumer decision-making process towards fresh meat. *British Food Journal*, 102(7): 522–38.

Verbeke W (2009). Market differentiation potential of country-of-origin, quality and traceability labeling. *The Estey Centre Journal of International Law and Trade Policy*, 10(1): 20–35.

Vermeir I & Verbeke W (2006). Sustainable food consumption: exploring the consumer 'attitude–behavioral intention' gap. *Journal of Agricultural and Environmental Ethics*, 19(2): 169–94.

Vermeir I & Verbeke W (2008). Sustainable food consumption among young adults in Belgium: theory of planned behaviour and the role of confidence and values. *Ecological Economics*, 64(3): 542–53.

Vitell S (2003). Consumer ethics research: review, synthesis and suggestions for the future. *Journal of Business Ethics*, 43(1–2): 33–47.

Wicks D (2011). Silence and denial in everyday life – the case of animal suffering. *Animals*, 1: 186–99.

Williams N (2008). Affected ignorance and animal suffering: why our failure to debate factory farming puts us at moral risk. *Journal of Agricultural and Environmental Ethics*, 21(4): 371–84.

Willits FK & Luloff AE (1995). Urban residents' views of rurality and contacts with rural places. *Rural Sociology*, 60(3): 454–66.

Zhao Y & Wu S (2012). Willingness to pay: animal welfare and related influencing factors in China. *Journal of Applied Animal Welfare Science*, 14(2): 150–61.

12

A utilitarian argument against animal exploitation

Clare McCausland

Animal abolitionism calls for an end to the exploitation of nonhuman animals. When this is advocated within a rights-based framework a justification falls readily from Kantian-inspired injunctions against instrumentalisation. For utilitarianism, however, an in-principle argument against exploitation is more difficult to make. It is for this reason that prominent abolitionist Gary Francione has argued at length that utilitarianism is not up to the task of anything more than merely regulating animal exploitation (Francione & Garner 2010, 8–11; Francione 1996, 61). According to utilitarianism, it follows that where animal exploitation maximises utility it should be endorsed rather than rejected. The view espoused by Peter Singer (1993, 67) that medical experimentation may in some instances be acceptable accords with this approach and has been strongly criticised by abolitionist rights theorists, and held to be representative of the relationship between welfarist and rights-based approaches to animal protection (cf. Francione 1996, 49ff).

In this chapter I show that supporters of utilitarianism can make an abolitionist argument against exploitation. The argument shows that the animal exploitation they oppose always sets back interests and never brings about a greater distribution of happiness than not exploiting animals. I start by identifying the kinds of animal exploitation unanimously rejected by abolitionists as instances of commercialised exploitation. This limits the scope of the required utilitarian argument. I ask whether the legal status of animals as property consistently results in less overall utility before considering economic arguments in greater detail. I show that while commodifying animals can produce a certain amount of utility for animal exploiters, in current market conditions it can only do so at the expense of an ever larger loss of utility for the animals they exploit.

Abolitionist arguments against exploitation

Abolitionists are divided in their attitudes towards the non-commercial use of animals. The morality of human interactions with companion animals is open to debate, for example the ownership of a companion animal offends some opponents of the property paradigm (Perz 2007; Dunayer 2004). Francione, on the other hand, has conducted a long-running social media campaign for the adoption of cats and dogs from New York City animal shelters (@garylfrancione). We can therefore surmise that he views his conception of abolitionism as consistent with the adoption of companion animals who would otherwise be euthanised, perhaps on the understanding that it is a lifesaving and therefore morally permissible (and laudable) action.

Opinion is similarly divided on the topic of zoos and sanctuaries. Using animals for the purpose of human entertainment is easily associated with privileging human interests over animal interests. Yet where the primary purpose of modern facilities is increasingly conservation or wildlife rescue the picture is less clear-cut. Here we find animal interests may be genuinely prioritised insofar as human benefit is restricted. Sanctuaries that both care for and exhibit rescued agricultural animals present similar difficulties.

Despite these contested areas of debate in animal use, what abolitionist approaches have in common is not a specific account of animal rights, but opposition to systematic and industrialised animal exploitation, regardless of the benefits which may follow. Abolitionists consistently oppose the farming of animals for food whether in concentrated or free-range facilities; the use of animals for pharmaceutical, scientific and cosmetic testing; the breeding and selling of animals as pets; and the commercial use of animals for entertainment, such as in rodeos and circuses. Furthermore, in all these activities, in addition to the privileging of human over nonhuman interests, the product of animal exploitation is bought and sold for profit. In capitalist societies it is these commercial institutions that abolitionists wish to see abolished.

Abolitionist arguments against systematised animal exploitation can draw on different foundations. Tom Regan's (1983) theory of animal rights identifies an individual's inherent value as an experiencing subject of a life that must be respected and on the basis of this, ought not to be instrumentalised. A relatively sophisticated level of cognition is key to Regan's conception of a subject of a life (1983, 78). Others, such as Evelyn Pluhar (1995), have developed rights theories underpinned by an animal's autonomy or agency. Pluhar (1995) draws specifically on Alan Gewirth's human rights theory and points to recognition on the part of autonomous agents that interfering with autonomy is profoundly detrimental to agency. This recognition compels moral agents to universalise the demand that autonomy be inviolate, which entails that instrumentalisation – and therefore exploitation – are fundamentally wrong. On the basis of her arguments Pluhar (1995, 256–58) suggests at the very least the set of rights holders will include mammals and most likely birds.

Utilitarians will not find these arguments very helpful; they refer to structural elements of a moral theory which is incompatible with their own. While utilitarianism might recognise moral status on the basis of our shared sentience with animals or recognise autonomy as an important interest and contributing factor to an individual's wellbeing, an in-principle rejection of exploitation cannot be defended on these grounds.

Francione's (1996; 2010) very popular brand of abolitionism focuses on the status of animals as property. People who subscribe to this view consider the property status of animals as a major societal obstruction standing in the way of recognising the moral status of animals. Abolishing the property status of animals is therefore a strong focus of attention for many animal protection scholars (Petrie 2009; Sankoff 2009a & 2009b; Dunayer 2007 & 2004; cf. Wise 2000). Abolitionists also consider property status in opposition to another of the key tenets of the position, namely that 'all sentient beings are equal for the purpose of not being used as resources' (Francione & Garner 2010, 5). We can therefore begin to assess whether utilitarianism can support a strong anti-exploitation position by looking to see whether it can argue against the property status of animals. Significantly, Francione (2010,

186) is of the view that if utilitarianism can make this claim, then it can at least defend a limited version of abolitionism. He explicitly leaves it to Peter Singer to do so, however, and since he has issued this challenge, no-one as far as I am aware has taken it up.

Property status

Francione (2010) advances the claim that there is only one right, and that is the right not to be property. His argument for this position proceeds along these lines: the right not to be property is held by anyone who has moral status. To have moral status, an individual must be sentient; that is, they must be capable of experiencing pain or pleasure. Francione applies the principle of equal consideration of interests to everyone with moral interests, including animals, and argues that since we do not treat other humans instrumentally, we ought not to treat other animals instrumentally. We should not, therefore, own them, or consider them property.

The property status of animals under law is – as was the property status of humans under law – structurally defective in his view. It wrongfully instrumentalises animals, allowing them to be used solely for human purposes, and is a guaranteed source of inequality and economic harm. Further, Francione (2010, 48–59) argues, it is a core belief of the abolitionist position that merely reforming animal regulation cannot be effective; abolitionism is in opposition to regulation. Almost no measure of improvements to welfare legislation can rectify this situation because as long as animals are property the inequality is structural.

For present purposes we do not need to assess the merits of this argument. I wish instead to consider the role which the property status of animals plays for abolitionism and its relationship to the exploitation of animals. Is this something which utilitarians can also utilise, as Francione (2010) suggests they might, in an effort to join the abolitionist cause?

Legal status

Francione (2010) identifies two types of argument against property: a legal argument and an economic argument. Legally, property owners have a prima facie right to exclusive use of their property. While this must be qualified, as will shortly be discussed, we might nevertheless think that this structural inequality leads to inequity in the distribution of utility between owners and those who are owned. The ownership relationship might point to a guaranteed flow of utility in the direction of property owners and away from those who are classified as property and who are also capable of experiencing utility; that is, from nonhuman animals. This structural inequality, therefore, might indicate that the ownership relationship itself can lead to assigning those who may be owned a lower status in society.

Francione (2010) also argues that the ownership relationship leads to a lower social status. It is not clear, however, that ownership and the legal rights to use that it conveys, necessarily leads to a lower social status for those who are owned and therefore a morally problematic state of affairs. The reasons for this are twofold. Firstly, legal restrictions upon ownership occur in myriad ways. The owner of a car must obey the road rules, the owner of a building must obey heritage and zoning laws that may significantly restrict their ability to modify their building or land, and the owner of an animal is subjected to animal welfare legislation. Extensive welfare legislation as well as outright bans on certain activities such as cock-fighting, types of animal testing, certain kinds of pet ownership and so forth, already place restrictions on animal owners.

Furthermore, property ownership is not a straightforward matter of having a right to exclusive use of something. AM Honoré (cited in Cochrane 2009, 427) notes that ownership confers multiple rights and duties, rather than a single right to exclusive use, and identifies eleven standard relations which an owner can bear in relation to their property.

Cochrane (2009, 428) asks: 'once we start breaking up the concept in this way, we are left to wonder whether there is anything left of it'. He also notes that despite the potential 'disintegration' of the concept of ownership, its existence conveys at least some information about the power relationship between the owner and the one owned. He suggests it tells us that owners have a certain priority over others in terms of the various rights and duties ownership conveys. This inequality should not be forgotten in the debate. The property relation is therefore harmful because of the structural inequality it engenders. If only humans may own property and this right defines their elevated status, and if nonhumans are defined by their inability to own property and by the fact that they can be owned, their interests can never be accorded the respect they deserve. It is from this distinction that the inevitable privileging of human over nonhuman interests stems.

Nevertheless, the potential for someone to abuse another under legal protection is no guarantee of abuse itself, at least where that specific entitlement cannot do any harm. This highlights an important difference between human and most nonhuman animals. A helpful comparison here might be between companion animals and the status of women a few generations ago. Both are or were the subject of ownership and indisputably of a lower social status. Yet even if one is fortunate to live in a benevolent household, the mere status of being owned does harm to a human being in a way that has no effect on a dog or a cat, who has no comparable interests in not being owned. As with all comparisons of interests between humans and nonhumans it is prudent to leave open the possibility of exceptional cases, and note that some animals may be cognitively sophisticated enough to recognise and resent a power imbalance, but we can be confident the loss of utility will not be as great as it is in other humans with equal relevant capacities to those who own them. For most nonhuman animals, if someone has a legal right to do them great harm, this is not evidence that they will.

Many laws exist which no longer reflect widespread moral norms and behaviour. For example, the exclusion of Indigenous Australians in the Commonwealth of Australia Constitution was widely recognised as a law at divergence from common moral views when the constitution was amended following the 1967 referendum. Yet Indigenous people had some measure of entitlements within every Australian state prior to the referendum, which did not subsequently improve either their rights of citizenship or welfare rights. Rather, the overwhelming success of the referendum, in which over 90 percent of the population voted for the ability of the Commonwealth to legislate for Indigenous people, is taken to be of symbolic significance to the people affected by the decision (Attwood et al. 2007). The symbolic importance of the decision and its meaning for Indigenous peoples cannot be disregarded. The example is intended to show that both a legal state of affairs can be at significant variance from the moral views of a population, and that changing this state of affairs may be of great importance primarily only to those who can appreciate its symbolic meaning. It is therefore plausible to suggest that the inequality represented by legal ownership rights of nonhuman animals is compatible with maximal utility. Property is therefore not a condition under which systemised exploitation should be rejected by utilitarianism.

Commodification

Marxian theory

If we do not think that legal property status alone guarantees a utility loss that should rule out animal exploitation, there is another way of thinking about the effects of this status which is relevant to the debate. In addition to the legal argument, Francione (2010) makes an economic argument. He writes: 'It is important to understand that animals are property. They are economic commodities; they have market value' (Francione & Garner 2010, 27). It is on the commodification of animals in this environment that I now want to focus. A focus on commodification reveals a Marxian inspiration for grounds on which utilitarianism might reject systematised animal exploitation. The claim is that the profit motive and imperative for increased productivity in a competitive market lead to one or two scenarios: the exploitation of more animals, leading to more intensified processing; and/or reduced costs, leading to a reduction in welfare through lower quality of confinement, veterinary care, staff and so on.

A growing body of literature points to capitalism as the explanation for increased animal suffering, and I suggest that Francione and the abolitionist/anti-reformist movement could be considered part of this literature, even if Francione does not explicitly advocate Marxism (Francione & Garner 2010, 228) or disavow the institution of private property altogether. Rather, he claims that animals should not be considered as items of property within the existing institution.

Ryan Gunderson (2011, 2) examined the Marxian literature and points to what he calls 'capital's blind drive for self-expansion and self-accumulation' as the cause of increased animal suffering. He points specifically to the shift of animal exploitation from one of use-value to exchange-value. Use-value is a Marxian term that describes the situation when something is valued according to its usefulness to the market. That which has use-value meets the needs of the person who produces it or their community (E Mandel cited in Gunderson 2011, 3).

There is no suggestion that according use-value to animals is harmful to their interests, or that use is inherently wrongful or inherently causes suffering. The next step in the process of commodification according to Marx, however, is the shift to assigning exchange-value to objects (Marx in Gunderson 2011, 3). This is when a commodity produced for the purpose of exchange on the market is valued in terms of that exchange. The producer of commodities no longer lives directly on the products of their own labour; on the contrary they can live only if they 'get rid of' these products (Mandel in Gunderson 2011, 3). People rear animals not for the purpose of meeting the needs of the community but for the purpose of accumulating capital.[1]

The effects of growth leading to unprofitability might be connected with Marx's (1999) argument that there is a tendency in any capitalist system for the rate of profit to decline. This argument rests on the relationship between necessarily increasing investment in 'constant capital', fixed assets and raw materials relative to the decreasing value of variable capital or (human) labour. The rate of profit declines because it is the result of dividing the value of surplus labour (the value of labour after its costs have been factored in) by the investment in the total means of production (constant capital plus variable capital). There is no avoiding the cost of further investment in constant capital: fixed assets decline in value and additional raw materials are required. The argu-

---

1   Marx (1859) argues that this transformation from use-value to commodity begins with the process of bartering, within which use-value still has primacy. It is only when a surplus is created – more than is needed for consumption – that what is valued for use comes to be valued for exchange. Since what is highly valued for use is frequently exchanged between communities, it is no surprise that these use-values serve as the first money. The words 'capital', 'cattle' and 'chattel' have a common source.

ment shows, therefore, that the profit motive exerts pressure on labour costs, resulting in detrimental effects on workers.

Marx (1999) considered money spent on livestock an investment in constant capital, either as a fixed asset or a raw material, depending on the use to which the animals were put. It might therefore be thought that animals are the beneficiaries of investment, and that we ought not to consider them in the same way we do human workers who play a very different role in production. Perhaps we ought therefore to consider that capitalism has the potential to benefit animals through constant investment in animal capital, all the while hurting human labourers via decreased labour investment and surplus labour value.

Animal welfare

In an animal-production system, money is spent on animals though not necessarily on the welfare of animals. Rather, it is spent buying, maintaining and using the animals. This need to invest in more animals can be interpreted as a need to increase the level of output, either by increasing the size of the entire operation or to increase the level of intensification by maintaining the same level of production and processing animals more rapidly and efficiently. If we believe that small-scale farming guarantees the highest overall utility gains for animals, an increase in the size of farms to accommodate increasingly more animals will be to the disadvantage of these animals if other investments in labour (that is, people to tend to animals' welfare needs) are not also made. If the alternative is to maintain the size of farms and the number of people looking after animals, but to increase the number of animals processed within them over time – in other words increasing the intensification of animal confinement – that too represents another way to decrease utility.

Humane farmers invest in welfare-enhancing technologies. Larger cages, for example, allow birds to perch where previously they could not. This, however, increases production costs. Animal welfare-enhancing technology may also fall under the category of constant capital, but importantly only to the extent that it furthers the means of production. If humanely treated animals are the product being sold, enhancement technology will be one of the ongoing investments a farmer must make. Marxian analysis tells us that these investments will affect the ability of humane farmers to profit if they do not simultaneously increase their productivity: as the value of the surplus labour declines, so declines the profitability of the company. Large-scale government subsidisation of animal agricultural industries suggests these pressures are already in place. Thus, the only way to maintain the surplus while continuing to invest in welfare-enhancing technologies will be to pay workers less, or to reduce the amount of time workers spend with the animals throughout the production process for the same level of output.

We have seen in practice what happens when farmers significantly reduce their investment in human labour: labour conditions at many intensive animal-production environments are particularly poor. Witness for example the repeated workplace investigations at even free-range poultry farms like those producing for the Lilydale brand in Australia, where employees have alleged they have been 'working under poor pay, harassment and fear' (Barlow & McClymont 2010). The effects of treating workers badly on animal welfare has also not gone unnoticed, even if it is sometimes dismissed by those whose primary focus is human rights (Whyte 2011).

Gunderson (2011) details what currently happens when producers breed animals and produce animal products for ever-declining profits. The account is familiar and disturbing. It is not dissimilar to horrific accounts of visible animal suffering in the 19th century which led to the development of early animal welfare legislation in the UK (cf. O'Sullivan 2011, 140–58), to the description of animal lives on farms and in laboratories in the 1970s (Singer 1995); or to the extensive accounts of institutionalised animal suffering offered by Francione and Garner (2010). The increasingly confined and painful lives of broiler chickens, egg-laying hens, and pigs and cows are discussed at great length at every point in animal-protection literature.

I consider one example here: the situation of free-range egg-laying hens. What might have begun as the backyard sale of eggs in welfare-rich situations has developed into to an extensive commercialised system of production. For example, the Australian Egg Corporation Limited (AECL) and Australian Poultry Industries Association (APIA) argued recently (if so far unsuccessfully) for 20 000 birds per hectare to be considered free-range (Carney 2012; Australian Competition and Consumer Commission 2012). Given the choice, chickens live in small groups of four to five (Page & Dawkins 1997, 23), and in smaller-size groups aggressive behaviour diminishes once birds establish a clear hierarchy. This is something that is not possible within the very large group sizes preferred by modern free-range producers. That chickens become aggressive towards each other is only one of many animal welfare problems that may result when chickens are forced to live in such close proximity.

'Free-range' is a term that is not currently defined by regulation in Australia (Australian Consumers Association 2012), so the size of farms from which eggs bearing this description is sold is unknown. In the US 'free-range' means that producers must demonstrate to the US Food Safety and Inspection Service that the birds have been allowed access to the outdoors. But, as Singer and Mason (2006, 92–93) discovered, access to the outdoors, which the United States Department of Agriculture also requires for hens laying eggs that are labelled as 'certified humane', may be interpreted as applying to the edges of a vast flock of tens of thousands birds, rather than to individual birds within a shed. These birds may in practical terms have very limited, if any, access to the outdoors.

Making money from animal welfare prompts the first objection to the claim that profitability always results in reduced animal utility. I argue, however, that animals still suffer even in best-case scenarios such as the production of free-range eggs and humanely farmed meat. For a farmer whose product is better-treated animals, it might be argued that the incentive to reduce welfare via reduced costs or by increasing the number of animals processed conflicts with something essential about their business. Rather, if the producers of better-treated animals are in competition with each other, the profit/growth motive should lead to greater welfare protections driven by consumer demand, rather than more harm. If this is the case, the related claim can be made that humane farming brings well-off animals into existence that would otherwise never have lived: an argument described by Matheny and Chan (2005) as the 'logic of the larder'.

To respond to this claim we can ask whether there is anything inherent in farming in a commercial environment – even free-range farming – which results in a net utility loss. We might first note that even if the term 'free-range' were robustly defined, the production of free-range eggs, as well as eggs produced by caged birds, results in a distinct and inevitable kind of 'wastage': the birth of male chicks. As unprofitable by-products of the egg-production industry, male chicks are killed within a couple of days of hatching. The male offspring of dairy cows face a similar fate. While some are killed for their flesh, 'bobby calves', as they are known, are mostly killed within days of their birth; the financial incentive for farmers to treat these young calves well is accordingly weak.

There is certainly a range of responses available to utilitarianism concerning the harm of death in nonhuman animals. Some find killing animals to be a comparatively minor moral transgression on the grounds that only humans seem to have the sort of capacities that can ground an interest in continuing to live (eg Singer 1979). Others provide a utilitarian argument that death is harmful on the basis of the foregone utility (eg Bradley 2009; Feldman 1991) that can equally be applied to animals. Despite the scope for disagreement, the likelihood of harm done to vast numbers of young male animals killed in these industries cannot be neglected in the utilitarian calculus.

The treatment of young male animals in the egg and dairy industries also responds to the logic-of-the-larder argument, where it can be demonstrated that a large number of the animals brought into existence experience no utility gains at all. Further to this point, Matheny and Chan (2005, 584ff) cite evidence to show that the number of animals killed or prevented from coming into existence through the habitat destruction involved in intensive animal agriculture is greater than the number brought into existence by agricultural breeding itself. If their calculations are correct, the argument against animal breeding can be made on the basis of number of lives lost alone. While this assessment of the number of lives lived and lost might speak against breeding, it does not address the possibility of inevitable harm being done to animals once they are alive.

This is a contested area of research, and a deep examination of the Marxian literature is beyond the scope of this chapter. The increasing harm done to animals in modern agricultural settings is not a new insight made by Marxian analysts. Nevertheless, the argument is compelling: the demand for growth and the process of commodification leads to a situation where exploiters must invest either in more animals or more intense means of production, or invest less in the people involved in the process of killing. We may not think that capitalism has failed, but its effects on nonhuman animals have certainly been striking.

Exploiter utility

Marx developed his analysis in the context of the Industrial Revolution and considered the harms inflicted by capitalism on human workers. From a utilitarian perspective these need to be balanced against the benefits accrued – not only by capitalists, but by human society as a whole. For example, while today many workers are almost certainly harmed in the production of mobile phone handsets (Litzinger 2013; Ayers 2012; Ngai & Chan 2012), mobile phones have brought an enormous benefit to humans worldwide that might be difficult to replicate with alternative technology. At the end of 2011 there were almost six billion mobile phone subscriptions, reaching 79 percent of the developing world alone (International Telecommunication Union 2011, 2).

A criticism made of utilitarianism by the advocates of rights-based approaches to animal ethics is that they must consider also the utility that comes to animal exploiters (eg Regan 1983, 220f). Given the extent of the suffering caused in the process of raising and killing animals, there is arguably something offensive about moral philosophers stopping to consider the pleasure people experience in eating animal flesh. But to disregard the numbers and the moral impact of aggregating large amounts of individual suffering is to disregard another powerful moral argument.

Over 50 billion animals are bred and killed each year for human consumption (Food and Agriculture Organization of the United Nations 2012). This figure does not include fish, who are also sentient; between one and three trillion fish are killed each year (Mood 2010, 72). Seven billion humans benefit from the suffering and death of these animals. These benefits are widespread and their pervasive effects on the human economy mean that an immediate cessation of activity that exploits animals would not be viable (cf. Frey 1983, 197–206). With the above ratios in mind, even if humans derived intensely pleasurable experiences from animal use, it is hard to see how these pleasures would outweigh the animal suffering involved in their production. It is important to note that comparable benefits of the kind that animal exploitation brings to humans can be generated elsewhere.

A shift from our current system of animal exploitation to a world economy that does not rely on this harm would be nothing less than revolutionary. Humans need to consider that food does not only come from animals and that jobs do exist outside of slaughterhouses, tanneries and circuses. A significant period of adjustment and perceived loss would need to be taken into account while people acclimatise themselves to new diets and new jobs, and in practice this period will not be months or years, but decades or centuries. Yet there is a compelling case to make that in the long term the abolition of this practice will produce no less utility for humans and substantially reduce suffering for the exploited animals.

Animals who cannot suffer

Utilitarianism seeks to maximise pleasure and to minimise suffering, and the argument against systematised animal exploitation rests on the likely fact that it fails to produce a greater overall amount of pleasure. One possible solution to this problem might therefore be to exploit animals without causing suffering. If that can be achieved it will bring only benefits to the exploiters just discussed. What if humans were able to breed and exploit animals that cannot feel pain? Douglas Adams (1980) writes of the horror and discomfort experienced by dinner guests faced with a cow inviting them to eat of his rump and shoulder, or perhaps a casserole made of him. 'Don't worry', the cow assures the reluctant dinner guests, 'I'll be very humane' (1980, 92–94).

Yet the possibility of eating non-sentient animals is not as fictional as it might first seem. People have adapted readily to drinking milk from non-animal sources, lending plausibility to the likely success of the slowly growing in-vitro meat movement, which has been investigated by scientists for at least a decade. There are now also people seeking to breed non-sentient animals as an alternate pathway to cruelty-free meat and dairy industries. Adam Shriver (2009) presents a challenge to utilitarian arguments against animal exploitation, citing recent research into genetic engineering. If we can breed animals with a significantly reduced or completely eliminated capacity to experience pain, he argues, we should not only replace animals currently farmed with such modified creatures, but utilitarians should revise any objections they have against animal exploitation all together.

Shriver (2009, 118) looks particularly at studies undertaken on genetic manipulation in mice. Mice with particular enzymes or gene expressions removed or 'knocked out' are still capable of sensory pain but show a reduction in the typical affective response to painful stimuli. That is, they are not as averse to pain in the way they normally would be. This provides evidence that they 'mind less', or are not as bothered by the prospect of a repeat occurrence of painful stimuli as they normally would be. This leads Shriver to speculate about the possibility of manipulating similar genes in the mammals we farm for food that are currently subjected to intense sensory and affective pain.

Shriver suggests that such gene manipulation might be effective for animals such as veal calves and pigs raised in sow stalls that, with their limited ability to move around, would be less likely to benefit from the warning signal that pain represents. He also cites studies on similar parts of the brain which have stopped mother mammals from responding to the cries of their young (2009, 119). This leads him to suggest that the current suffering inherent to the dairy industry might be prevented by breeding appropriately modified cows. Shriver (2009, 121) notes some limitations of this research. In particular the genes studied have only a limited effect on the affective response. The area of the brain studied by the authors he cites, the anterior cingulate cortex, is involved in a wide range of experiences we associate with emotional suffering. Nevertheless, our understanding of how the full range of human and animal suffering is represented in the brain is far from complete.

Further risks are posed by producing these genetically modified animals. As Shriver notes (2009, 122), one of the results of breeding animals less capable of experiencing some kinds of pain might be that they are treated more callously by those responsible for their care. In such a scenario, genetically modified animals might then experience much greater suffering of the kind they are capable of experiencing independently of the genetic modification. The production of such animals therefore carries a risk of causing even greater harm. Eliminating some sources of pain might reduce some of the harm of exploitation, but may result in an increase in pain from other sources.

It is also not entirely clear that Shriver's challenge presents a problem only for utilitarians looking to argue for abolitionism. Earlier we noted that Francione's conception of animal rights rests on animal capacity for sentience, and sentience certainly plays a role in Regan's formulation of being a 'subject-of-a-life' (Regan 1983, 243). If non-sentient animals can be bred, would they be entitled to rights? Sentience does not represent all of the qualities Regan values and, as noted, other rights scholars such as Pluhar (1995) ground animal rights in autonomy. It might also be possible that genetically modified animals that experience less affective pain are still autonomous, but if we construe autonomy as second-order desire, this is not obviously the case. It is reasonable to understand even first-order desire as an affective state of the kind that might be eliminated by the sort of genetic modification Shriver cites. If that is the case, then arguably non-sentient animals would not hold rights to freedom from exploitation on any understanding of rights.

The appropriate response to this emerging research is caution. Both supporters of utilitarianism and rights theorists have reason to be suspicious that grave moral harms to animals can currently be avoided with the techniques developed so far. It may be the case that these techniques are the precursor to more sophisticated technologies that prevent animals (and from then, humans) from experiencing significant suffering. If so, a utilitarian theorist would need to determine the level of suffering each animal is likely to experience and multiply this by the number of animals that will likely benefit from the technology, together with the benefits accrued to human exploiters. If it seems that a greater overall amount of utility will be produced, then utilitarianism should simply condone the exploitation of such modified creatures. Animal rights advocates must consider the extent to which an ideal technology inhibits a creature's capacity for autonomy. If only non-autonomous, insentient creatures are being exploited, then they, too, have new questions to answer.

Conclusion

One of the fundamental differences between the rights-based approach to animal protection and the utilitarian approach is the response to animal exploitation. This chapter asked whether utilitarianism could endorse the claim made by animal rights scholars that the commercial exploitation of animals ought to be abolished. In the current global market and in situations that are likely to arise in the future, I argue they can. While an argument grounded in the legal status of animals as property was not evidence of guaranteed lower utility, a Marxian analysis of commodification shed light on the negative effects that occur when those who are exploited for profit are only valued in terms of that profit. The replaceable utility obtained by the comparatively small number of humans who consume animal products cannot outweigh the suffering experienced by the substantially larger number of animals exploited under these increasingly worse conditions. While the possibility of producing animals that cannot suffer remains on the horizon as a panacea for all earthly suffering, until this solution is complete, it may raise more problems than it solves. The conclusion, therefore, for utilitarians to draw – and for those who believe in rights to draw – is that the exploitation of animals for profit is an activity to be avoided.

Acknowledgements

I am very grateful to Neil Levy, Ned Dobos and Siobhan O'Sullivan for their critical feedback and encouragement, to the attendants of the philosophy staff seminar at Charles Sturt University at which an earlier version of this paper was presented, and to two anonymous reviewers whose insightful comments helped me to make significant improvements to the style and substance of this paper.

Works cited

Adams D (1980). *The restaurant at the end of the universe*. London: Pan Books.
Attwood B, Markus A, Edwards D & Schilling K (1997). *The 1967 referendum, or, when Aborigines didn't get the vote*. Canberra: Australian Institute of Aboriginal and Torres Strait Islander Studies.
Australian Competition and Consumer Commission (2012). Australian Poultry Industry Assoc. CTM 2012. [Online] Available: www.accc.gov.au/content/index.phtml/itemId/1060311 [Accessed 26 February 2014].

Ayers CJ (2012). The international trade in conflict minerals: coltan. *Critical Perspectives on International Business*, 8(2): 178–93.

Barlow K & McClymont A (2010). Poultry producer's workers claim intimidation. *ABC News*, 22 October. [Online] Available: www.abc.net.au/news/2010-10-21/poultry-producers-workers-claim-intimidation/2307364 [Accessed 12 October 2012].

Bradley B (2009). *Well-being and death*. Oxford: Clarendon Press.

Carney M (2012,). 'Free range' definition under scrutiny. *ABC News*, 31 May. [Online] Available: www.abc.net.au/news/2012-05-31/free-range-definition-under-scrutiny/4043558 [Accessed 12 October 2012].

Cochrane A (2009). Ownership and justice for animals. *Utilitas*, 21(4): 424–42.

Dunayer J (2007). Advancing animal rights: a response to 'antispeciesism,' critique of Gary Francione's work, and discussion of speciesism. *Journal of Animal Law*, 3: 1–32.

Dunayer J (2004). *Speciesism*. New York: Lantern Books.

Feldman F (1991). Some puzzles about the evil of death. *The Philosophical Review*, C(2): 205–27.

Food and Agriculture Organization of the United Nations (2012). FAOSTAT. [Online] Available: http://faostat.fao.org/site/573/default.aspx [Accessed 27 February 2014].

Francione GL (1996). *Rain without thunder: the ideology of the animal rights movement*. Philadelphia: Temple University Press.

Francione GL & Garner R (2010). *The animal rights debate*. New York: Columbia University Press.

Frey RG (1983). *Rights, killing, and suffering: moral vegetarianism and applied ethics*. Oxford: Blackwell.

Gunderson R (2011). From cattle to capital: exchange value, animal commodification, and barbarism. *Critical Sociology*: 1–17. doi: 10.1177/0896920511421031.

International Telecommunication Union (2011). The world in 2011 – ICT facts and figures. [Online] Available: www.itu.int/ITU-D/ict/facts/2011/material/ICTFactsFigures2011.pdf [Accessed 12 October 2012].

Litzinger R (2013). The labor question in China: Apple and beyond. *South Atlantic Quarterly*, 112(1): 172–78 .

Matheny G & Chan KMA (2005). Human diets and animal welfare: the illogic of the larder. *Journal of Agricultural and Environmental Ethics*, 18: 579–94.

Marx K (1999 [1867]). *Capital*. Edited by F Engels, translated by S Moore & E Aveling. Moscow: Progress Publishers. [Online] Available: www.marxists.org/archive/marx/works/1867-c1/ [Accessed 12 October 2012].

Marx K (1970 [1859]). *A contribution to the critique of political economy*. Translated by NI Stone. New York: International Publishers.

Mood A (2010). Worse things happen at sea: the welfare of wild-caught fish. [Online] Available: www.fishcount.org.uk/published/standard/ fishcountfullrptSR.pdf [Accessed 12 October 2012].

Ngai P & Chan J (2012). Global capital, the state, and Chinese workers. The Foxconn experience. *Modern China*, 38(4): 383–410.

O'Sullivan S (2011). *Animals, equality and democracy.* Basingstoke: Palgrave Macmillan.

Page M & Dawkins MS (1997). Peck orders and group size in laying hens: 'futures contracts' for non-aggression. *Behavioural Processes*, 40: 13–25.

Perz J (2007). Adulterating animal rights: Joan Dunayer's 'Advancing animal rights' refuted. *Journal of Animal Law & Ethics*, 2: 123–72.

Petrie L (2009). Companion animals: valuation and treatment in human society. In P Sankoff & S White (eds), *Animal law in Australasia: a new dialogue* (pp57–78). Leichhardt: The Federation Press.

Pluhar EB (1995). *Beyond prejudice.* Durham, NC: Duke University Press.

Regan T (1983). *The case for animal rights.* London: Routledge & Kegan Paul.

Sankoff P (2009a). Animal law: a subject in search of scholarship. In P Sankoff & S White (eds). *Animal Law in Australasia: A New Dialogue* (pp389–400). Leichhardt: The Federation Press.

Sankoff P (2009b). The welfare paradigm: making the world a better place for animals? In P Sankoff & S White (eds), *Animal law in Australasia: a new dialogue* (pp7–34). Leichhardt: The Federation Press.

Shriver A (2009). Knocking out pain in livestock: can technology succeed where morality has stalled? *Neuroethics*, 2: 115–24.

Singer P (1995). *Animal liberation.* 2nd edn. London: Pimlico.

Singer P (1993). *Practical ethics.* 2nd edn. Cambridge: Cambridge University Press.

Singer P (1979). Killing humans and killing animals. *Inquiry*, 22(1): 145–56.

Singer P & Mason J (2006). *The ethics of what we eat.* Melbourne: Text Publishing.

Whyte D (2011). Rare victory for workers whose dignity was cut to the bone. *Sydney Morning Herald*, 25 November. [Online] Available: www.smh.com.au/ opinion/politics/ rare-victory-for-workers-whose-dignity-was-cut-to-the-bone-20111124-1nwwj.html [Accessed 12 October 2012].

Wise SM (2000). *Rattling the cage: towards legal rights for animals.* Cambridge: Perseus Publishing.

13

Emotions in animals

Nicky McGrath and Clive Phillips

The experience of feeling an emotion can be taken for granted in humans. Everything we do is coloured by some kind of emotion. We can be paralysed by fear or overwhelmed with happiness. These familiar emotions manifest themselves in our daily lives. Observing the behaviour of animals close to us, we may believe they are experiencing similar feelings. The challenge for science, however, is to categorise or label these feelings, and to be confident in our interpretation.

Public concern for animals focuses on nonhuman animals that are considered to be sentient or feeling beings, and this is reflected in legislation worldwide. The UK Farm Animal Welfare Council (FAWC) urges that the future for farm animal welfare ought to be based on every animal having 'a life worth living' as a minimum, and, if possible, 'a good life' (FAWC 2009, 14–16).

Animal welfare relates to the experiences of animals under human care. More specifically, these experiences can contribute to good or poor welfare. In order to establish whether, from an animal point of view, they are having 'a good life', it is important to assess their emotions. As Burman et al. (2008, 330) state, 'An understanding of emotional states is critical because it is the presumed existence of such states that underlies the public concerns about animal welfare.'

Scientists face difficulty when trying to provide empirical proof of feelings in animals who cannot express themselves verbally. Inferring emotions in animals from their behaviour has often been the only possible approach. Until recently, research on farm animal welfare has mainly been linked to stress (Boissy et al. 2007a, 37), however, progress in neuroscience allows detection of areas in the brain associated with the processing of certain discrete emotions (Berridge 2003, 27–43; Panksepp 1998). The homologies between humans and animals mean that neuroscience may help to determine exactly what feelings animals experience. The literature provides a truly complex picture but, increasingly, research is looking at ways of measuring positive welfare in animals, rather than just stress or negative emotional states (Boissy et al. 2007b, 376–77). New methods of measuring emotions are being developed which can serve to complement and validate existing methods. This chapter discusses these methods and their limitations, and highlights the importance of reliable measurements of emotions in animals for the assessment of welfare.

What constitutes an emotion?

Philosophers, neuroscientists, psychologists and biologists have all contributed definitions of emotion. Across disciplines, many differentiate between basic emotions, such as happiness and fear, and those regarded as more complex emotions, like guilt or embarrassment (Ekman 1999). Some include pleasure and pain as emotions (Dawkins 1998, 883–87), while others describe emotions as being separate from these affective states (Ortony & Turner 1990, 328), or from motivational states like hunger and thirst (Hacker 1999, 4). Further distinctions are made between primary emotions, which are the instinctive responses to stimuli, and secondary emotions, which are more complex and follow primary emotions (Plutchik 2001, 348–50; Panksepp 1998).

In relation to animals, Darwin (1872) was the first to describe basic emotions. He cited anger, happiness, sadness, disgust, fear and surprise as universal core emotions. This view of emotions as discrete entities was later supported by Ekman (1999) who maintained that individual emotions have universal antecedent causes as well as other characteristic hallmarks.

In contrast to describing emotions as discrete entities, some theorists suggest that the study of emotions should focus on different dimensions such as positive or negative valence, which has benefit for ascribing effects on welfare, or high and low arousal (Barrett 1998, 595–97). A mechanistic approach is championed by neuroscientists, who study the neural circuits and neurochemicals associated with emotions (LeDoux 1994, 68–75). In addition, some scientists look at emotions in terms of components, with most describing physical and behavioural components as common to all emotions (Boissy et al. 2007a, 37; Désiré et al. 2002, 166).

There is a danger that classifications of emotions into discrete entities or different valence neglects the complexity of their varying role in different species and their degree of positive/negative valence or level of arousal. In addition, neuroscientific studies, as well as offering little to those wishing to make inferences about welfare, are likely to require invasive surgery which itself affects animals' welfare.

Boissy et al. (2007a, 37) describe an emotion as 'an intense but short-lived affective response to an event, which is associated with specific body changes'. This definition includes physical and behavioural elements, but does not mention any cognitive or subjective components that humans may associate with 'feelings' (Désiré et al. 2002, 166). Broom and Fraser (2007, 331) define a feeling as

a brain construct involving at least perceptual awareness which is associated with a life regulating system, is recognisable by the individual when it recurs and may change behaviour or act as a reinforcer in learning.

Damasio (2003, 28) differentiates feelings from emotions saying that for the purposes of neuroscience, emotions are the physical reactions to certain stimuli, while feelings occur after we become aware of the physical changes.

Experiences describe direct observation of, or participation in, events as a basis of knowledge. These are precursors for physical sensations, which are in turn antecedents for emotions, following the perceptual and cognitive processing which complex animals undertake in order to become aware of emotions. Recently, experiences have been proposed as the most reliable measure of welfare, although their complexity hinders attempts to categorise their valence (Phillips 2009; Bracke et al. 1999, 311–19).

The evidence for a subjective component in animal emotions is keenly debated in the literature. The positivism and behaviourism of the 20th century focused on measuring only what can be observed, disregarding any subjective feelings of animals. Behaviourists also regarded emotions as simply epiphenomena, with no function for behavioural control (Panksepp 2005, 41). Their legacy still haunts contemporary science, with a dichotomy of views evident on whether animals can feel their emotions or whether they are simply automatic responses.

Why do emotions exist?

Emotions are generally said to be about something, or to have an 'object' (Hacker 1999, 12). The object can be any external or internal stimulus which elicits an emotional response. In evolutionary terms, emotions are adaptive responses that enable animals to solve problems they would be unable to with simple automatic reflexes. Emotions, therefore, offer behavioural flexibility. Most emotions are short term and intense, mobilising physical responses which enable animals to avoid harm or approach resources that will improve their fitness (Boissy et al. 2007b, 377; Rolls 2000, 177). In simple terms, they provide an adaptive benefit (Bekoff 2007).

Rolls (2000, 179–81) cites nine purposes of emotions: to prepare the body for action, to allow flexibility in behavioural response, to motivate, to communicate to others, to facilitate social bonding, to mediate cognitive evaluation of events or memories, to facilitate storage of memories, to help achieve goals by directing motivation towards them and to trigger recall of memories. All of these could be important contributors to an animal's welfare state.

Panksepp (1998) agrees with only some of these functions, suggesting that emotional systems coordinate both behavioural and physiological processes. He asserts that subjective feeling states may function to channel behaviour and help learning, thus aiding decision-making, and also helping to anticipate events, especially those which may improve or threaten fitness. Damasio et al. (2000, 84) also think that emotions aid in decision-making, as demonstrated by the irrational behaviour of brain-damaged patients, who lack the function of certain brain areas associated with emotions and are unable to strategically plan for the future.

The argument that emotions fulfil a social role also has wide support (Bekoff 2007; Ekman 1999; Humphrey 1976, 303–17). Darwin argued that emotions evolved as tools to facilitate social bonding in animal groups (cited in Bekoff 2007). He regarded facial expressions as important social communication devices. Ekman (1999) agrees and suggests that facial expressions have an informative role for conspecifics. Easily differentiated expressions reflecting anger, fear, happiness and disgust, have been documented in humans, primates and some other animals (Bennett & Hacker 2005, 29–52) and may provide an inner window into their welfare state. In higher mammals, however, emotions may not always provide all the information about their welfare state, which may have to be inferred from other characteristics, such as physiological indicators.

William James (1884) controversially postulated that physiological reactions induce emotional feelings (James–Lange theory). Arguments against this soon emerged. Separation of the viscera and the brain did not diminish emotional response (Dalgleish 2004, 582), and less invasive modern experiments have confirmed this. The Cannon–Bard theory subsequently argued that the hypothalamus is the centre for controlling emotional responses to stimuli, but that the neocortex provides a level of control (known as top-down control). Removing the cortical area of the brain results in uninhibited emotional responses (Dalgleish 2004, 583).

There is a definite bias towards a mechanistic approach in neuroscience. Barnard et al. (2007, 6–7) propose Interactive Cognitive Subsystems (ICS) in which animals' emotions are directly linked to motor action and have nothing to do with thought. They do concede, however, that great apes may possess a subsystem that allows a form of internal mental representation to occur simultaneously with action. This possibility may mean that the welfare state of great apes may not simply be ascertained from physiological signs and behaviour.

Damasio (1998, 84) posits his somatic marker hypothesis which suggests that associations between reinforcing stimuli induce a physiological affective state. A network of markers influences decision-making in situations which may involve conflicting messages or stimuli, and directs the animal's attention towards advantageous options, thereby simplifying the decision-making process. In this way, emotions play a critical role in the ability to make fast, rational decisions in complex and uncertain situations.

The relationship between cognition and emotion is another debated topic. Lazarus and Smith (1990, 615–22) believe cognition is a precondition for emotions, while Zajonc (1984, 117–22) says emotion and cognition are separate and affect can be generated without a prior cognitive process. Lazarus and Smith (1990, 615–22) suggest all emotions are subject to cognitive appraisal, of which there are two types: automatic, unreflective and either unconscious or preconscious; or deliberate and conscious. The appraisal generates action relevant to specific environmental conditions. Emotions expand response flexibility and reduce the stimulus specificity that characterises reflexes.

Tooby and Cosmides (1990, 375–424) believe any appraisal is automatic and based on ancestral evidence rather than upon present-day environmental cues. This does not allow for flexible evaluation of current situations and therefore reduces the utility of some emotional responses. Panksepp (1998) agrees that raw reflexes are fundamental to emotions, citing the fact that few humans are able to overcome intense feelings with cognitive skill. Zajonc (1984, 117–22) also cautions against accepting the subordination of emotions to cognitive control and says this would greatly reduce their adaptive value.

Indeed, some behavioural responses, such as jumping away from what is perceived to be a snake before realising it is a stick, are innate in humans and nonhuman animals (Panksepp 1998). They do not require cognitive input. Conditioned fear is also largely involuntary. LeDoux (1994, 69) believes this type of fear can never be erased as it remains dormant in the amygdala, the part of the brain associated with processing emotions and memory storage.

Another view on the relationship between cognition and emotion suggests that emotions mediate cognition. Boissy et al. (2007a, 40) posit that emotions can bias the processing of information coming from the environment, resulting in changes in attention, memory and judgment of a situation. For example, anxiety functions to direct attention towards threatening stimuli, and memories formed during emotionally charged situations are more readily remembered. This has adaptive value for organisms in terms of survival. LaBar and Cabeza (2006, 55) also believe emotion affects memory by focusing attention on emotionally salient details of events. The main effects are beneficial, although chronic stress (as seen in anxiety) can have deleterious effects.

Panksepp (1998) proposes feedback systems between higher cortical areas with cognitive control, subcortical basic emotional systems and physiological and behavioural changes in the body. He believes that secondary emotions, as opposed to the primary emotional responses, allow animals to think in terms of perceptual changes. He also discusses the apparently unique ability of humans to have thoughts about thoughts, which requires 'linguistic-symbolic transformation of simple thoughts and remembered experiences' (Panksepp 2005, 32).

Emotions: implications for animal welfare

The aforegoing discussion indicates that the close connection between emotions and cognitive processes may enable us to infer one from the other, helping our understanding of the welfare state of animals. However, the question remains as to whether only animals that feel the psychological components of emotions, and by inference have cognitive abilities, can be said to have a welfare state. If we accept that experiences form the basis of welfare assessment, then clearly an animal without cognition can have a welfare state, even though it is not aware of it, in just the same way that an animal may harbour a disease without being aware that it is doing so. We may further argue that cognition is not required for attribution of a welfare state using either Broom's (1986, 524) definition of welfare as an animal's state as it attempts to cope with its environment or Duncan's (1993, 8–14) consideration that welfare depends on what animals feel.

Consciousness in animals has been debated for over 350 years, since Descartes. Emotions and consciousness are often considered to be intrinsically connected (Izard 2009, 2). Cabanac (1999, 8) says any animal that has emotion possesses consciousness. Consciousness, in his terms, refers to 'presence of a mental space' (Cabanac et al. 2009, 271). However Rolls (2000, 177–234) posits that conscious experience requires linguistic symbolisation of behavioural states, and LeDoux (1994, 75) does not believe animals can have conscious affective experience.

The existence of language in humans is often used as a differentiating factor in determining whether animals have consciousness or can experience higher-level emotions (Wynne 2004). Ekman (1999), although admitting there is no evidence that humans experience different emotions from nonhuman animals, suggests that the type of experience of those emotions is changed by the ability to represent them in words. A Cartesian mechanistic approach favoured by behaviourists is no longer the norm in science, however doubts about the conscious experience of animals still linger (Carruthers 2005, 19; Dennett 1995, 691). The deadlock may be broken by accepting that there are various degrees of conscious awareness (Young 1994; Griffin 1976).

Measuring emotions in nonhuman animals

If we accept that emotions play a role in attributing welfare states, it is important to be able to accurately record and analyse emotions. Emotions therefore have an important function, and to have a function, Dawkins (1993) maintains, there must also be an effect. Dawkins argues that emotions are necessary for reinforcement learning (1998, 886–87). Hence, if a stimulus is reinforcing it must be related to an emotion. Therefore, operant or instrumental conditioning experiments in the vein of Pavlov (1928), Skinner (1938) or Thorndike (1927) would be a good starting place to measure emotions.

The laboratory setting is, however, unlikely to be suitable for revealing all types and facets of emotions in animals. Using a componential approach reveals other methods. Désiré et al. (2002, 166) describe emotions as having three elements: behavioural (posture or activities), autonomic (visceral and endocrine responses) and subjective (psychological emotional experience or feeling). Evidence can, therefore, also be found in behavioural, physiological, and cognitive signs. Combining these methods with neuroscience produces acceptable evidence to suggest an animal is experiencing emotions (Panksepp 1998). Some concurrent measurements may, however, provide conflicting information, for example in the assessment of stereotypical behaviour (Duncan 2005, 484).

Looking at situations that elicit emotions has also been suggested as a way of distinguishing discrete emotions (Dantzer 2001; Scherer

1997, 902–22). However, situations may elicit assorted responses at different times (Mendl et al. 2010b, 2895–904). Qualitative behavioural methods championed by Wemelsfelder (2007, 27–30) could enhance this method, as measuring gross reactions in animals could give a more holistic view of what the animal is feeling.

Other approaches include using expert knowledge of caretakers who provide their own opinion on what animals are feeling. Qualitative approaches such as free-choice profiling (FCP) rely on good knowledge and experience of species and individual animals (Wemelsfelder 2007, 30). Questionnaire responses have been shown to correlate well with behavioural measures in some cases, offering another complementary route to establishing what emotions animals are feeling (Morris et al. 2008, 3–18).

Neuroscientific methods

Affective neuroscience is the study of the neural mechanisms of emotion. Panksepp (1998) states that there are many homologous neural structures in the brains of all mammals, especially in the subcortical areas. Techniques used in affective neuroscience include electrical stimulation of the brain (ESB), creating brain lesions, lateralisation studies, pharmacological activation or inhibition of neurotransmitter receptors, monitoring of neuronal firing patterns using microelectrodes, and gene expression measurement.

By manipulating the brain using these techniques, and correlating the outcomes with observation of approach and avoidance behaviours, Panksepp (1998) has established the existence of five basic emotions (fear, anger, separation distress, play and seeking) in the mammalian brain. These emotions, he says, create 'action tendencies', flexible responses to situations where animals need to learn strategies for survival. Further proof that these emotions exist in nonhumans may become evident if key neurochemistries and brain activation patterns in animals are correlated with predicted emotional responses in humans (Panksepp 2005, 36–37).

Current human brain studies are done using non-invasive techniques like electroencephalography (EEG), positron emission topography (PET) and functional magnetic resonance imaging (fMRI) which measure neuron action potentials and blood flow in the brain.

However, these techniques have limitations in terms of measuring emotional functions, especially in subcortical areas that are densely packed with neurons. They are also expensive (Berridge 2003, 40–41).

Some scientists believe that the subcortical and cortical areas are responsible for different types of emotions. Many posit that subcortical areas are responsible for innate emotions, with some suggesting that these function as reflex responses to external stimuli (Berridge 2003, 40–41; Panksepp 1998; LeDoux 1994, 75). Most reserve the cortical areas (the neocortex) as responsible for control of higher emotions, requiring some kind of cognitive or even conscious evaluation (Berridge 2003, 40–41; Damasio et al. 2000, 1049–53; Rolls 2000, 190–91).

Evidence in neuroscience about emotions in nonhuman animals relates to responses to pleasant or unpleasant stimuli. However, there is some disagreement among scientists as to whether or not a pleasant stimulus can induce an emotion. The argument centres on whether affect is actually emotion. LeDoux's research has apparently been able to isolate fear as a discrete emotion affecting the amygdala (LeDoux 1994, 70–73), and brain imaging has revealed that the amygdala in humans is affected by facial expressions showing fear, sadness or anger (Berridge 2003, 31).

Rolls (2000, 178) emphasises that, to understand the brain mechanisms of emotions, scientists need to understand how the brain decodes the reinforcement value of primary reinforcers. He rejects the James–Lange theory and Damasio's somatic marker hypothesis, saying that peripheral factors do not take part in decisions. Although neuroscience has a place in measuring animal emotions, invasive techniques and inducing negative emotions clearly have ethical implications.

Physiological methods

Ekman (1999) believes there are physiological correlates with emotions which prepare the organism to respond effectively. Indeed, much research into nonhuman animal emotions involves taking physiological measurements in combination with observing the animal's behaviour. Methods include measuring autonomic responses such as heart rate, respiration rate and electrodermal responses (Reefmann et al. 2009, 651–59; Désiré et al. 2002, 167). Other methods measure arousal of the hypothalamic–pituitary–adrenal axis (HPA) by analysing plasma cortisol levels, which can indicate a stress response, and arousal of the sympathetic–adrenal axis (SA) as manifested through changes in adrenaline and noradrenaline secretion. The immune response is also sometimes used as a measure (Laudenslager et al. 1990, 247–64).

Kober et al. (2008, 999) state that there has been little success in establishing 'unique signatures' for emotions like anger, fear and sadness. The issue here is being able to differentiate discrete emotions. Heart rate, often used as an indicator of stress, can also increase with sexual activity or due to anticipation of reward or punishment. Hence, it is an indicator of arousal (Paul et al. 2005, 475).

The sampling method can also confound results. For example, stress can be induced by restraint during blood sampling (Slaughter et al. 2002, 408), and care has to be taken with ceiling and circadian rhythm effects on glucocorticoid levels (Molony & Kent 1997, 271). Using physiological measurements alone is not, therefore, a valid way of measuring emotions in animals. Correlations between multiple physiological indicators and behavioural responses would allow better interpretation of emotions (Paul et al. 2005, 480; Dantzer 2001, E1–9).

Behavioural observations: quantitative methods

Behavioural measurements include observing the postures, locomotion, facial expressions and vocalisations of animals. Place preference tests, consumer demand tests and approach and avoidance studies are used to differentiate emotions. To be reliable indicators, behaviours should occur consistently in different contexts (Paul et al. 2005, 473–74). Facial expression is thought to be an indicator of fear in horses (Leiner & Fendt 2011, 108–09). Panksepp (1998) disagrees that facial expression is a useful measure, but advocates measures of vocalisations and has famously suggested that rats can laugh (Panksepp 2003, 383). Other studies suggest vocalisations indicate a wide variety of emotions such as fear (Seyfarth & Cheney 2003, 32–51; Leiner & Fendt 2011, 106–08), separation distress (Panksepp 2003, 379; Weary & Fraser 1995, 1048; Laudenslager et al. 1990, 252–54) and compassion (Douglas-Hamilton et al. 2006). Postures and locomotion are widely used as behavioural measures, and have been suggested to indicate grief in chimpanzees (Cronin et al. 2011, 4–5), and both positively and negatively valenced emotions in sheep (Reefmann et al. 2009, 651–59).

Quantitative behavioural measurements used concurrently with other methods can infer emotions (Panksepp 1998). Biplot analysis to explore principal components in a suite of measured behaviours can assist in identification of related behaviours (Kohler & Luniak 2005, 208–23), and comparison with human behaviours associated with known emotions can provide evidence of their existence in animals. There is room for development in this field; for example Boissy et al. (2007b, 309) believe that further research into farm animal vocalisations could reveal certain 'markers' for positive emotions.

Although behavioural measurements are quantifiable and repeatable, they have limitations when assessing emotions. It can be difficult to attribute emotions without using humans as a model. Facial expressions can be ambiguous; for example the 'grin' exhibited by primates could indicate either fear or anger (Wemelsfelder 2007, 28), not the amusement or pleasure associated with the behaviour in humans. Animals may be motivated to approach predators, perhaps to gather information, or in cooperative defence (Graw & Manser 2007, 507–17; Walling et al. 2004, 164–70); therefore, approach cannot always be interpreted as reward-seeking. Behaviours may exhibit plasticity, varying in intensity, duration or both. Controlled experiments may not enable expression of natural behaviour or emotion (Dawkins 2008, 941; Bekoff 2000, 868). Lastly, responses are frequently species-specific (Burman et al. 2008, 330), and fine movements are recorded out of context of the whole animal (Wemelsfelder 1997, 77–83).

Behavioural observations: qualitative methods

Paul et al. (2005, 473) believe interpretation of behaviour is problematic because significance is determined by the observer. However Bekoff (2000, 861) says looking at the real-life situation is paramount, as emotions evolve in contexts. Wemelsfelder (1997, 83) believes that the subjective experience of nonhuman animals is accessible and requires a change in both method and mindset. Rather than quantifying individual fragmented behaviours, she urges scientists to look at the animal as a whole, as an agent or a 'behaver'. Her method of free-choice profiling involves a set of observers generating their own descriptions of behaviour, and then rating each individual animal against the descriptions. The method has been applied to many animal species, including pigs (Wemelsfelder et al. 2009, 477–84), dairy cows (Rousing & Wemelsfelder 2006, 40–53), dogs (Walker et al. 2010, 75–84) and horses (Napolitano et al. 2008, 342–54). The utility of this approach has been acknowledged, especially as a method for assessing positive emotions (Boissy et al. 2007b, 390), and Wemelsfelder and Lawrence (2001, 21–25) advocate using this method to validate other types of behavioural measurement.

Animal behaviour can be misleading. Subjective qualitative assessment of behaviours may introduce a personal bias, and all assessors may misinterpret the same cues. Paul et al. (2005, 469–91) do not mention qualitative techniques for measuring emotional processes indicating a reluctance to concede qualitative methods as valid. Boissy et al. (2007b, 390), despite suggesting that qualitative assessment may be a good measure of positive emotions in animals, point out that it is difficult to validate this method of assessment as there is no benchmark.

Cognitive behavioural methods

According to Boissy et al. (2007a, 39), sheep (and presumably other animals including chickens [Zimmerman et al. 2011, 569–77]) can develop expectations. Any inconsistency between these expectations and reality can produce negative emotional arousal, which is substantiated by behavioural agitation and increased heart rate (Boissy et al. 2007a, 39; Veissier et al. 2001, 2580–93). These results indicate that it is the animal's representation of an event, rather than the event itself, that governs its reaction (Boissy et al. 2007a, 38). Cognitive processes should, therefore, be taken into account to assess emotional experiences of animals.

As discussed previously, some scientists believe that emotions are subject to cognitive appraisal. Most appraisal theories outline criteria of evaluation which are used by humans and animals, either automatically or consciously, and which result in discrete emotions (Frijda et al. 1989, 212–28; Scherer 1988, 89–126). Désiré et al. (2002, 176) propose to measure behavioural and physiological responses of farm animals, whilst controlling for individual and pairs of appraisal criteria, such as suddenness, novelty, predictability and intrinsic pleasantness. Scherer (1988, 89–126) developed these criteria for his 'Stimulus Evaluation Checks', which an organism uses sequentially. The complexity of the checks performed is dependent on the age and phylogenetic position of the animal.

An experiment to assess emotions in animals using this method would aim to expose animals to one criterion at first, such as novelty, suddenness or predictability, followed by combining criteria. Behavioural and physiological responses would be measured and related directly to the evaluation criteria to establish emotional outcomes (Paul et al. 2005, 480; Désiré et al. 2002, 176). Thus, these experiments could provide a framework to detect a range of emotions by looking at different eliciting situations (Paul et al. 2005, 480). Individual differences may also be uncovered, which may have implications for breeds or strains used in certain farming practices.

Assessment of underlying mood or affective state is a relatively new approach to measuring welfare in animals (Harding et al. 2004). Repeated arousal of negative emotion has been demonstrated to have a deleterious effect on animals, resulting in states indicative of despair or depression. For example, fear and frustration in mice induced by inescapable stressful situations, or separation of mother and infant macaques, lead to agitated behaviour (Cryan et al. 2005, 571–625; Kaufman & Stynes 1978, 71–75). Chronic activation of these emotional states appears to result in learned helplessness, or depression (Boccia et al. 2007, 69).

In humans, research has shown that people in a negative affective state are more sensitive to loss or failure (Burman et al. 2008, 330) and anxious people make more negative judgments in their interpretation of ambiguous stimuli (Eysenck et al. 1991, 144–50). Harding et al. (2004, 312) developed a technique to assess 'cognitive bias' in animals to determine whether animals were experiencing longer-term negative emotional states such as depression. The technique tests animals assumed to be under a negative emotional state by presenting them with positive, negative and ambiguous stimuli. Experiments have appeared to successfully show a negative bias in rats from unpredictable housing conditions (Harding et al. 2004, 312), rats in unenriched housing (Burman et al. 2008, 331) and dogs showing separation-related behaviours (Mendl et al. 2010a, R839). In addition, short-term manipulation of light intensity in rats' environments seemed to produce a cognitive bias which indicated anxiety (Burman et al. 2009, 350).

Animals may be exposed to specific stimuli that affect their appetite for positive stimuli, such as food, without affecting their emotional state. Often the two are intrinsically linked, but situations may be conceived in which the wrong conclusions could be drawn. For example, supposing a researcher wanted to test whether animals subjected to movement were experiencing an emotion similar to depression in humans. If they used the animals' willingness to advance for a food reward in an ambivalent situation, a reduction in this desire might be interpreted as evidence of depression, whereas it might simply indicate nausea following the movement. Thus there are problems of interpretation that are common with other measures.

A second issue is that cognitive bias experiments are more suited to controlled environments which may not facilitate expression of all emotions (Bekoff 2000, 868). Also, individual animals may perceive a situation differently, which may result in inconsistent behavioural and physiological responses (Mendl et al. 2010b, 2900). In addition, although a cognitive bias test may suggest a negative emotional state, such as depression, it may not always reveal the eliciting cause.

Core affective space framework methods

Recently, Mendl et al. (2010a, 2895–904) proposed a dimensional approach which moves away from simply assessing discrete emotions. Their core affective framework locates discrete emotions within a two dimensional space characterised by levels of valence (positive or negative) and arousal (high or low). These components of discrete emotions serve to prioritise action. Discrete emotions arising from events can have short-term effects on the overall mood state (core affect), and also a cumulative effect on longer-term mood. For example, persistent exposure to negative events may result in a high-arousal, negative mood state. Mood state, therefore, reflects past experiences, and also constantly changes according to individual experiences. Mood states also then have a reciprocal effect on discrete emotions and decision-making, as discussed above.

Measuring emotions based on this framework involves making predictions about the types of situations that will generate a particular affective state, predicting the types of decisions animals will make in certain states and assessing underlying mood states by identifying discrete emotions associated with these states. This approach has the potential to identify positive affective states and discrete emotions (Mendl et al. 2010a, 2895–904). Humans may not categorise all emotions as positive or negative, however, and may transition quickly between them. Therefore, animals may also experience intangible emotional states.

Stereotypies: a means to assess negative emotional states?

Stereotypies are repetitive behaviours which do not have an obvious goal (Mason 1991, S57). They are thought to be indicative of poor welfare (Redbo 1998, 273–78), and therefore, in many captive situations, environmental enrichment measures are introduced (Fernandez et al. 2008, 200–12; Cooper et al. 1996, 244). Mason and Latham (2004, S57–69) suggest that stereotypies that become habit-like may not actually be evidence for poor current welfare and that they may function as a coping mechanism which induces pleasure. Others believe that they have a largely physiological function, such as oral stereotypies enhancing production of salivary buffers in stressed animals with acid-induced gastrointestinal discomfort (Moeller et al. 2008, 85–90).

Anecdotal methods: field studies

Qualitative, narrative approaches are often taken when measuring emotions. Fraser (2009, 113) calls this the 'alternative paradigm', suggesting that scientists need to infer emotion in animals to make credible interpretations of behaviour. Darwin (1872) and Romanes (1883) were among the first to compile accounts of emotions in animals. Other scientists provide rich narrative descriptions of the emotional experiences of some animals. Jane Goodall is one such scientist who has captured public attention with her work on chimpanzees (Goodall 1986). Joyce Poole and Carole Moss have extensively studied elephants in the wild (Bekoff 2007). Sue Savage-Rumbaugh is well known for reports on the behaviour, language and emotions of primates (Masson & McCarthy 1996), and Marc Bekoff has a long experience of fieldwork measuring emotions in animals (Bekoff 2007).

Bekoff (2007, 121) defends his method of assessment by claiming that 'the plural of anecdotes is data'. Emphasising that emotions evolve in a context (Bekoff 2000, 86), he suggests that it is possible to create experiments which closely simulate anecdotal situations (Bekoff 2007). Others discuss problems that arise from scientific reluctance to accept anecdotal data as evidence for emotions (Masson & McCarthy 1996).

Anecdotal methods: owner and caretaker reports

As expert opinion and experience is key to some qualitative methods of measuring emotions in animals, it appears there is room for evidence from owners and caretakers, as well as the qualitative researcher assessments discussed above. Indeed Fraser (2009, 109) says 'Many of the important questions about animal welfare arise when people, drawing on their "everyday" understanding of animals, are concerned about animals' affective states.' Morris et al. (2008, 16) found owners' reports of emotions in their pets both coherent and consistent with the ability to differentiate between discrete emotions such as affection, joy, sadness and jealousy. However, anecdotal accounts of animal emotions are often subject to accusations of anthropomorphism. Fraser (1999, 185) cautions against 'gratuitous speculation', and suggests authors make errors in their attempts to describe animal subjective experiences as being the same as human ones. Heyes (1993, 177–79) states that anecdotes are an inadequate means of attributing mental states to animals.

Anthropomorphism as a method

There are many proponents of anthropomorphism as a useful tool for establishing whether and how animals feel emotions. Attribution of human characteristics to animals may help infer how an animal is feeling and Panksepp (2003, 382) believes that anthropormorphic reasoning may be appropriate if animal behaviour can consistently be linked to affective state.

Bekoff (2007, 123) says anthropomorphism is a 'linguistic tool to make the thoughts and feelings of other animals accessible to humans', and agrees with Serpell (2003, 86) that it comes naturally to humans. Burghardt (2007, 136–38) advocates 'critical anthropomorphism' as the basis of ethological research. Bateson (1991, 835) believes that good science depends on people putting themselves inside the minds of animals and that this is key to the measurement of animal emotion. He maintains that ethologists can, through projection of human emotions, predict behaviours such as attack or escape in animals (Bateson 1991, 836).

Caution is essential, as human judgments can be erroneous and misinterpretation could lead to assumptions which negatively affect animal welfare. If certain emotions include a psychological component, for example depression or grief, then scientific evidence for this emotion in the animals being observed is also necessary to avoid anthropomorphic projections. Ultimately the efficacy of this method depends on the similarity between human and animal behaviours in the emotions they relate to, the ability of observers to understand animals' behaviours, and observers' understanding of the connection between human emotions and the behaviours that they display. Although it would not be permitted to make such an assumption in the modern age of critical reasoning, Darwin (1859) believed that the similarities in human and animal emotions were so obvious that they did not need description. Kennedy (1992), however, criticises the field of cognitive ethology for being too anthropomorphic. Wynne (2007, 125) maintains that anthropomorphism does not allow for objective measurement and McGrath et al. (2013, 42) suggest that the attribution of emotions to animals extends beyond scientific evidence.

Why measure emotions in animals? Implications for animal welfare

Dawkins (1998, 838–88) and Bekoff (2007) both believe that providing evidence of emotions in animals is crucial to improving the welfare of animals under our care. Increasingly, animal welfare studies look at whether animals are having a good life. The animal must make a value judgment on what is good, which depends clearly on how it feels (Boissy et al. 2007a, 38). 'Telos', a term first used by Aristotle, is defined in terms of animals' needs and interests (Hauskeller 2005, 64). Comparing the concept of telos to recent definitions of animal welfare, for example, Dawkins' definition of animal welfare as 'Health and what animals want' (2008, 937), we can see striking similarities. Rollin (1998, 346) argues that 'the interests comprising the telos are plausibly what matters most to the animals'. If this is the case, understanding animal emotions may be the key to providing them with what they want.

Phillips (2009) distinguishes between animals' needs and desires, commenting that most animals are actually deficient in their needs.

In the laboratory setting, unhappy or stressed animals can confound experimental outcomes. The ability to assess and improve emotional states in this situation could decrease the number of animals used in experiments (Meunier 2006, 326–47). Panksepp (1998) believes measuring neural mechanisms for emotions in animals could potentially advance the study and treatment of affective disorders in humans; nevertheless, both Panksepp and LeDoux discuss the inability to obtain funding for research into emotions in animals (Panksepp 2005, 35). The legacy of behaviourism, which precluded any discussion or research into emotion, apparently lives on in behavioural neuroscience.

Conclusion

The difficulties of measuring emotions in animals stem from the inability to see into the workings of another being's mind. Animal behaviour may convince us that they have emotions, but equally, we may not easily infer emotions in animals we are not familiar with. The debate we introduce is not whether animals experience emotions, but rather what impact these subjective experiences have on their welfare. Although Bentham (1789, 238) famously said 'the question is not, can they reason, nor can they talk, but can they suffer?', science has advanced such that the question is 'can we know the extent to which they suffer?'

This chapter discussed the many methods available to scientists to provide evidence of emotions in animals. Used concurrently rather than individually, these methods may provide a more reliable understanding of animal welfare and deepen our interpretation of the priorities of complex animals in relation to their welfare. The current challenge in the face of major welfare issues is to acknowledge some utility in methods and apply them to real-life situations to improve the lives of animals under our care.

Works cited

Barnard PJ, Duke DJ, Byrne RW & Davidson I (2007). Differentiation in cognitive and emotional meanings: an evolutionary analysis. *Cognition and Emotion,* 21: 1155–83.

Barrett LF (1998). Discrete emotions or dimensions? The role of valence focus and arousal focus. *Cognition and Emotion,* 12: 579–99.

Bateson P (1991). Assessment of pain in animals. *Animal Behaviour,* 42: 827–39.

Bekoff M (2007). *The emotional lives of animals.* Novato, CA: New World Library.

Bekoff M (2000). Animal emotions – exploring passionate natures. *Bioscience,* 50: 861–70.

Bennett M & Hacker P (2005). Emotion and cortical-subcortical function: conceptual developments. *Progress in Neurobiology,* 75: 29–52.

Bentham J (1789). *Introduction to the principles of morals and legislation.* Oxford: Clarendon Press.

Berridge KC (2003). Comparing the emotional brains of humans and other animals. In RJ Davidson, KR Scherer & H Goldsmith (eds), *Handbook of affective sciences* (pp25–51). Oxford: Oxford University Press.

Boccia ML, Razzoli M, Vadlamudi SP, Trumbull W, Caleffie C & Pedersen CA (2007). Repeated long separations from pups produce depression-like behavior in rat mothers. *Psychoneuroendocrinology,* 32: 65–71.

Boissy A, Arnould C, Chaillou E, Désiré L, Duvaux-Ponter C, Greiveldinger L, Leterrier C, Richard S, Roussel S, Saint-Dizier H, Meunier-Salaun M, Valance D & Veissier I (2007a). Emotions and cognition: a new approach to animal welfare. *Animal Welfare,* 16: 37–43.

Boissy A, Manteuffel G, Jensen M, Moe R, Spruijt B, Keeling L, Winckler C, Forkman B, Dimitrov I & Langbein J (2007b). Assessment of positive emotions in animals to improve their welfare. *Physiology & Behavior,* 92: 375–97.

Bracke MBM, Spruijt BM & Metz JHM (1999). Overall animal welfare reviewed. Part 3: Welfare assessment based on needs and supported by expert opinion. *Netherlands Journal of Agricultural Science,* 47: 307–22.

Broom DM (1986). Indicators of poor welfare. *British Veterinary Journal,* 142: 524–26.

Broom DM & Fraser AF (2007). *Domestic animal behaviour and welfare.* Cambridge: CABI.

Burghardt G (2007). Critical anthropomorphism, uncritical anthropomorphism and naive nominalism. *Comparative Cognition and Behavior Reviews,* 2: 136–38.

Burman OHP, Parker RMA, Paul ES & Mendl M (2008). Sensitivity to reward loss as an indicator of animal emotion and welfare. *Biology Letters,* 4: 330–33.

Burman OHP, Parker RMA, Paul ES & Mendl MT (2009). Anxiety-induced cognitive bias in non-human animals. *Physiology & Behavior* 98: 345–50.

Cabanac M (1999). Emotion and phylogeny. *Japanese Journal of Physiology,* 49: 1–10.

Cabanac M, Cabanac AJ & Parent A (2009). The emergence of consciousness in phylogeny. *Behavioural Brain Research,* 198: 267–72.

Carruthers P (2005). Why the question of animal consciousness might not matter very much. *Philosophical Psychology,* 18: 83–102.

Cooper JJ, Odberg F & Nicol CJ (1996). Limitations on the effectiveness of environmental improvement in reducing stereotypic behaviour in bank voles (clethrionomys glareolus). *Applied Animal Behaviour Science,* 48: 237–48.

Cronin KA, van Leeuwen EJC, Mulenga IC & Bodamer MD (2011). Behavioral response of a chimpanzee mother toward her dead infant. *American Journal of Primatology,* 73: 415–21.

Cryan JF, Mombereau C & Vassout A (2005). The tail suspension test as a model for assessing antidepressant activity: review of pharmacological and genetic studies in mice. *Neuroscience & Biobehavioral Reviews,* 29: 571–625.

Dalgleish T (2004). The emotional brain. *Nature Reviews Neuroscience,* 5: 582–89.

Damasio AR (1998). Emotion in the perspective of an integrated nervous system. *Brain Research Reviews,* 26: 83–86.

Damasio AR (2003). *Looking for Spinoza: joy, sorrow and the feeling brain.* London: William Heinemann.

Damasio AR, Grabowski TJ, Bechara A, Damasio H, Ponto LLB, Parvizi J & Hichwa RD (2000). Subcortical and cortical brain activity during the feeling of self-generated emotions. *Nature Neuroscience,* 3: 1049–56.

Dantzer R (2001). Can we understand farm animal welfare without taking into account the issues of emotion and cognition? *Journal of Animal Science,* 79: 32.

Darwin C (1872). *The expression of the emotions in man and animals.* 3rd edn. Oxford: Oxford University Press.

Dawkins MS (2008). The science of animal suffering. *Ethology,* 114: 937–45.

Dawkins MS (1998). Animal minds and animal emotions. *American Zoologist,* 38: 7A.

Dawkins MS (1993). *Through our eyes only? The search for animal consciousness.* Oxford: WH Freeman.

Dennett DC (1995). Animal consciousness – what matters and why. *Social Research,* 62: 691–710.

Désiré L, Boissy A & Veissier I (2002). Emotions in farm animals: a new approach to animal welfare in applied ethology. *Behavioural Processes,* 60: 165–80.

Douglas-Hamilton I, Bhalla S, Wittemyer G & Vollrath F (2006). Behavioural reactions of elephants towards a dying and deceased matriarch. *Applied Animal Behaviour Science,* 100: 87–102.

Duncan IJH (1993). Welfare is to do with what animals feel. *Journal of Agricultural and Environmental Ethics,* 6 (Suppl 2): 8–14.

Duncan IJH (2005). Science-based assessment of animal welfare: farm animals. *Revue Scientifique et Technique OIE,* 24: 483–92.

Ekman P (1999). Basic emotions. In T Dalgleish & M Power (eds), *Handbook of cognition and emotion* (pp45–60). Sussex: Wiley & Sons.

Eysenck MW, Mogg K, May J, Richards A & Mathews A (1991). Bias in interpretation of ambiguous sentences related to threat in anxiety. *Journal of Abnormal Psychology,* 100: 144–50.

FAWC (2009). Farm animal welfare in great britain: past, present and future. [Online] Available: www.fawc.org.uk/pdf/ppf-report091012.pdf [Accessed 1 June 1 2011].

Fernandez LT, Bashaw MJ, Sartor RL, Bouwens NR & Maki TS (2008). Tongue twisters: feeding enrichment to reduce oral stereotypy in giraffe. *Zoo Biology,* 27: 200–12.

Fraser D (1999). Animal ethics and animal welfare science: bridging the two cultures. *Applied Animal Behaviour Science,* 65: 171–89.

Fraser D (2009). Animal behaviour, animal welfare and the scientific study of affect. *Applied Animal Behaviour Science,* 118: 108–17.

Frijda NH, Kuipers P & Ter Schure E (1989). Relations among emotions, appraisal, and emotional action readiness. *Journal of Personality and Social Psychology,* 57: 212–28.

Goodall J (1986). *The chimpanzees of Gombe: patterns of behaviour.* Cambridge, MA: Harvard University Press.

Graw B & Manser MB (2007). The function of mobbing in cooperative meerkats. *Animal Behaviour,* 74: 507–17.

Griffin DR (1976). *The question of animal awareness: evolutionary continuity of mental experience.* The New York: Rockefeller University Press.

Hacker PM (1999). The conceptual framework for the investigation of emotions. *International Review of Psychiatry,* 16: 199–208.

Harding EJ, Paul ES & Mendl M (2004). Animal behavior: cognitive bias and affective state. *Nature,* 427: 312–12.

Hauskeller M (2005). Telos: the revival of an aristotelian concept in present day ethics. *Inquiry: An Interdisciplinary Journal of Philosophy,* 48: 62–75.

Heyes CM (1993). Anecdotes, training, trapping and triangulating: do animals attribute mental states? *Animal Behaviour,* 46: 177–88.

Humphrey NK (1976). The social function of intellect. In PPG Bateson & RA Hinde (eds), *Growing points in ethology* (pp303–17). Cambridge: Cambridge University Press.

Izard CE (2009). Emotion theory and research: highlights, unanswered questions, and emerging issues. *Annual Review of Psychology,* 60: 1–25.

James W (1884). What is an emotion? *Mind,* 9:188–205.

Kaufman IC & Stynes AJ (1978). Depression can be induced in a bonnet macaque infant. *Psychosomatic Medicine,* 40: 71–75.

Kennedy JS (1992). *The new anthropomorphism.* Cambridge: Cambridge University Press,.

Kober H, Barrett LF, Joseph J, Bliss-Moreau E, Lindquist K & Wager TD (2008). Functional grouping and cortical–subcortical interactions in emotion: a meta-analysis of neuroimaging studies. *NeuroImage,* 42: 998–1031.

Kohler U & Luniak M (2005). Data inspection using biplots. *Stata Journal,* 5: 208–23.

LaBar KS & Cabeza R (2006). Cognitive neuroscience of emotional memory. *Nature Reviews Neuroscience,* 7: 54–64.

Laudenslager ML, Held PE, Boccia ML, Reite ML & Cohen JJ (1990). Behavioral and immunological consequences of brief mother-infant separation: a species comparison. *Developmental Psychobiology,* 23: 247–64.

Lazarus RS & Smith CA (1990). Emotion and adaptation. In LA Pervin (ed.), *Handbook of personality: theory and research* (pp609–37). New York: Guilford.

LeDoux J (1994). Emotion, memory and the brain. *Scientific American,* 270: 50–57.

Leiner L & Fendt M (2011). Behavioural fear and heart rate responses of horses after exposure to novel objects: effects of habituation. *Applied Animal Behaviour Science,* 131: 104–09.

Mason GJ (1991). Stereotypies: a critical review. *Animal Behaviour,* 41: 1015–37.

Mason GJ & Latham NR (2004). Can't stop, won't stop: is stereotypy a reliable animal welfare indicator? *Animal Welfare,* 13: S57–S69.

Masson J & McCarthy S (1996). *When elephants weep: the emotional lives of animals.* London: Vintage.

McGrath N, Walker J, Nilsson D & Phillips C (2013). Public attitudes towards grief in animals. *Animal Welfare,* 22: 33–47.

Mendl M, Brooks J, Basse C, Burman O, Paul E, Blackwell E & Casey R (2010a). Dogs showing separation-related behaviour exhibit a 'pessimistic' cognitive bias. *Current Biology,* 20: R839–R40.

Mendl M, Burman OHP & Paul ES (2010b). An integrative and functional framework for the study of animal emotion and mood. *Proceedings of the Royal Society,* B 277: 2895–904.

Meunier LD (2006). Selection, acclimation, training, and preparation of dogs for the research setting. *ILAR Journal*, 47: 326–47.

Moeller BA, McCall CA, Silverman SJ, McElhenney WH & Wendell H (2008). Estimation of saliva production in crib-biting and normal horses. *Journal of Equine Veterinary Science*, 28: 85–90.

Molony V & Kent JE (1997). Assessment of acute pain in farm animals using behavioral and physiological measurements. *Journal of Animal Science*, 75: 266–72.

Morris P, Doe C & Godsell E (2008). Secondary emotions in non-primate species? Behavioural reports and subjective claims by animal owners. *Cognition & Emotion*, 22: 3–20.

Napolitano F, De Rosa G, Braghieri A, Grasso F, Bordi A & Wemelsfelder F (2008). The qualitative assessment of responsiveness to environmental challenge in horses and ponies. *Applied Animal Behaviour Science*, 109: 342–54.

Ortony A & Turner TJ (1990). What's basic about basic emotions? *Psychological Review*, 97: 315–31.

Panksepp J (2005). Affective consciousness: core emotional feelings in animals and humans. *Consciousness and Cognition*, 14: 30–80.

Panksepp J (2003). Can anthropomorphic analyses of separation cries in other animals inform us about the emotional nature of social loss in humans? Comment on Blumberg and Sokoloff (2001). *Psychological Review*, 110: 376–88.

Panksepp J (1998). *Affective neuroscience: the foundations of human and animal emotions*. Oxford: Oxford University Press.

Paul ES, Harding EJ & Mendl M (2005). Measuring emotional processes in animals: the utility of a cognitive approach. *Neuroscience & Biobehavioral Reviews*, 29: 469–91.

Pavlov IP (1928). *Lectures on conditioned reflexes*. Edited and translated by WH Gantt. New York: International Publishers.

Phillips C (2009). *The welfare of animals: the silent majority*. Dordrecht: Springer.

Plutchik R (2001). The nature of emotions. *American Scientist*, 89: 344–50.

Redbo I (1998). Relations between oral stereotypies, open-field behavior, and pituitary-adrenal system in growing dairy cattle. *Physiology & Behavior*, 64: 273–78.

Reefmann N, Wechsler B & Gygax L (2009). Behavioural and physiological assessment of positive and negative emotion in sheep. *Animal Behaviour*, 78: 651–59.

Rollin BE (1998). On telos and genetic engineering. In A Holland & A Johnson (eds), *Animal biotechnology and ethics* (pp156–71). London: Chapman & Hall.

Rolls ET (2000). Precis of the brain and emotion. *Behavioral and Brain Sciences,* 23: 177–234.

Romanes G (1883). *Animal intelligence.* Washington DC: University Publications of America.

Rousing T & Wemelsfelder F (2006). Qualitative assessment of social behaviour of dairy cows housed in loose housing systems. *Applied Animal Behaviour Science,* 101: 40–53.

Scherer KR (1988). Criteria for emotion-antecedent appraisal: a review. In V Hamilton, GH Bower & NH Frijda (eds), *Cognitive perspectives on emotion and motivation* (pp89–126). Norwell: MA Kluwer Academic.

Scherer KR (1997). The role of culture in emotion-antecedent appraisal. *Journal of Personality and Social Psychology,* 73: 902–22.

Serpell JA (2003). Anthropomorphism and anthropomorphic selection – beyond the 'cute response'. *Society & Animals,* 11: 83–100.

Seyfarth RM & Cheney DL (2003). Meaning and emotion in animal vocalizations. *Annals of the New York Academy of Sciences,* 1000: 32–55.

Skinner BF (1938). *The behavior of organisms: an experimental analysis.* Oxford: Appleton-Century.

Slaughter MR, Birmingham JM, Patel B, Whelan GA, Krebs-Brown AJ, Hockings PD & Osborne JA (2002). Extended acclimatization is required to eliminate stress effects of periodic blood-sampling procedures on vasoactive hormones and blood volume in beagle dogs. *Laboratory Animals,* 36: 403–10.

Thorndike EL (1927). The law of effect. *The American Journal of Psychology,* 39: 212–22.

Tooby J & Cosmides L (1990). The past explains the present: emotional adaptations and the structure of ancestral environments. *Ethology Sociobiology* 11: 375–424.

Veissier I, Boissy A, dePassille AM, Rushen J, van Reenen CG, Roussel S, Andnason S & Pradel P (2001). Calves' responses to repeated social regrouping and relocation. *Journal of Animal Science,* 79: 2580–93.

Walker J, Dale A, Waran N, Clarke N, Farnworth M & Wemelsfelder F (2010). The assessment of emotional expression in dogs using a Free Choice Profiling methodology. *Animal Welfare,* 19: 75–84.

Walling CA, Dawnay N, Kazem AJN & Wright J (2004). Predator inspection behaviour in three-spined sticklebacks (*gasterosteus aculeatus*): body size, local predation pressure and cooperation. *Behavioral Ecology and Sociobiology,* 56: 164–70.

Weary DM & Fraser D (1995). Calling by domestic piglets: reliable signals of need? *Animal Behaviour,* 50: 1047–55.

Wemelsfelder F (2007). How animals communicate quality of life: the qualitative assessment of behaviour. *Animal Welfare,* 16: 25–31.

Wemelsfelder F (1997). The scientific validity of subjective concepts in models of animal welfare. *Applied Animal Behaviour Science,* 53: 75–88.

Wemelsfelder F & Lawrence AB (2001). Qualitative assessment of animal behaviour as an on-farm welfare-monitoring tool. *Acta Agriculturae Scandinavica, Section A – Animal Science,* 21–25.

Wemelsfelder F, Nevison I & Lawrence AB (2009). The effect of perceived environmental background on qualitative assessments of pig behaviour. *Animal Behaviour,* 78: 477–84.

Wynne CDL (2007). What are animals? Why anthropomorphism is still not a scientific approach to behavior. *Comparative Cognition and Behavior Reviews,* 2: 125–35.

Wynne CDL (2004). *Do animals think?* Princeton, NJ: Princeton University Press.

Young AW (1994). Neuropsychology of awareness. In A Revonsuo & M Kampinnen (eds), *Consciousness in philosophy and cognitive neuroscience* (pp173–204). Hillsdale: Erlbaum.

Zajonc RB (1984). On the primacy of affect. *American Psychologist,* 39: 117–23.

Zimmerman PH, Buijs SAF, Bolhuis JE & Keeling LJ (2011). Behaviour of domestic fowl in anticipation of positive and negative stimuli. *Animal Behaviour,* 81: 569–77.

Author biographies

Sally Borrell is secretary of the Australian Animal Studies Group and an associate editor of its new *Animal Studies Journal*. She is interested in literary representations of animals, particularly in terms of postcolonialism and post-humanism. She has recently published on JM Coetzee's *Disgrace* (in Manish Vyas [ed.], *Issues in Ethics and Animal Rights*). Sally wrote her PhD at Middlesex University in London addressing human–animal relations in postcolonial literature, and has a Master's degree from the University of Canterbury. She is an associate of the New Zealand Centre for Human–Animal Studies.

Jill Bough is a conjoint fellow of the School of Humanities and Social Sciences at the University of Newcastle. After 38 years as an educator and academic in England and Australia, she combined her lifelong passion for animals and her academic career by completing her PhD, From value to vermin: the donkey in Australia. In 2009, Jill was co-convenor of the Minding Animals conference, which was the third conference of the Animals and Society (Australia) Study Group and the first international conference dedicated to animal studies. Her particular interest is the use and representation of donkeys in human cultures. In 2011 she published *Donkey* with Reaktion Books, London.

Georgette Leah Burns has a background in cultural anthropology and has been working in the environmental sciences at Griffith University since 1996. The results of her longstanding research interest in tourism impacts were first published following a fieldtrip to Nepal in 1989. Her research has recently focused on the interactions between humans and wildlife, especially in nature-based tourism settings. Leah is author of numerous book chapters and journal articles on this topic, including the book *Dingoes, penguins and people: engaging anthropology to reconstruct the management of wildlife tourism interactions* (Lambert Academic Publishing, 2010). She is deputy coordinator of the Australia's Past and Present research cluster in Griffith's Environmental Futures Research Institute and is a former vice chairperson of the Australian Animal Studies Group.

Sandra Burr is an adjunct professional associate in the Faculty of Arts and Design at the University of Canberra where she also teaches creative writing and research methods. She belongs to the faculty writing research cluster, and is a member of the editorial panel for the online journal *Axon: Creative Explorations*. She writes reviews for several publications including the *AASG Bulletin, M/C Reviews* and *Text: Journal of Writing and Writing Courses*. Sandra has published in a number of scholarly and popular journals and is a member of the newly created Centre for Creative and Cultural Research. She has a PhD in creative writing and her research interests include human–animal relations and creative research methodology.

Simone Dennis is a senior lecturer in anthropology at the Australian National University, where she convenes the honours and foundational programs in anthropology and teaches the anthropology of human–animal relations and the anthropology of science. She is the author of three monographs dealing with various aspects of contemporary Australian life. Simone's research interests coalesce around anthropological theories of embodiment, politics and science. These interests have been explored in ethnographic work on Christmas Island shortly after the Tampa crisis, in work among Persian women migrants to Australia who have fled Iran in the past two decades, in research conducted in the technoscientific spaces of major Australian research laboratories in which mice and rats feature as animal models for human disease research, and in research which looks at cigarette smoking in urban Australian settings in this smoke-free era.

Sophie Fern is a writer and scientist currently based in Dunedin, New Zealand. She has published a number of books and articles on science for children, has worked in television documentary research and has taught both marine science and nonfiction writing at a university level. Her current research interests include the effective communication of science and the communication of journey and discovery in a scientific context.

Sally Healy earned a Bachelor of Biology at Griffith University then completed Honours in environmental psychology. Continuing her interest in human relationships with the natural world, Sally is currently undertaking a PhD in consumer behaviour and animal welfare, examining the connection between consumers and their food, in particular, animal-based products such as meat, dairy, and eggs. This research utilises both qualitative and quantitative methods to examine people's attitudes, knowledge, and concerns surrounding animal farming in Australia. With this research Sally hopes to add to understandings of the factors that support ethical consumerism and public awareness of animal farming. Sally has presented her work at several conferences including the multidisciplinary Minding Animals conference in the Netherlands in July 2012. She is the 2011 recipient of the RSPCA Australia Scholarship for Humane Animal Production Research.

Elizabeth Leane is an associate professor of English at the University of Tasmania, where she holds a research position split between the School of Humanities and the Institute for Marine and Antarctic Studies. She holds degrees in both literature and science. She is the author of *Reading popular physics* (Ashgate, 2007) and *Antarctica in fiction: imaginative narratives of the far south* (Cambridge University Press, 2012), and the co-editor of *Considering animals* (Ashgate Publishing Group, 2011) and *Imagining Antarctica: cultural perspectives on the southern continent* (Quintus Publishing, 2011). Her research has been published in journals such as *The Review of English Studies, Signs, Ariel, Anthrozoös* and *Polar Record*, and she is arts editor of the *Polar Journal*. In 2004 she travelled south on an Australian Antarctic Arts Fellowship.

Nicholas Malone is a lecturer in the Department of Anthropology at the University of Auckland. His research investigates the behaviour and ecology of endangered primates. Additionally, he is interested in the ethical implications of human and alloprimate interactions. His writing is informed by research experience in Indonesia and the Democratic Republic of the Congo.

Clare McCausland wrote a PhD on the relationship between ethical theories and animal protection movements at the University of Melbourne, where she is currently a member of the Human Rights & Animal Ethics Research Network. Clare works in academic governance and planning and is a committee member of the Australian Animal Studies Group. Her current research looks at animal autonomy and civil disobedience on behalf of animals.

Nicky McGrath graduated with a degree in Spanish and French from the University of Exeter in 1997 and then studied a Master's of Science in applied animal behaviour and animal welfare at the University of Edinburgh in 2011. She is undertaking research for her PhD on the vocalisations of chickens at the University of Queensland.

Kate Nash is a lecturer in the School of Media and Communication at the University of Leeds. Kate has a background in science communication and began her career as a producer at the Australian Broadcasting Corporation's radio science unit. Kate's research focuses on nonfiction media, particular documentary. She is currently looking at the emergence of interactive nonfiction media and teaches courses in journalism and multimedia. Kate is co-editor of *New documentary ecologies: emerging platforms, practices and discourses* (Palgrave Macmillan, 2014).

Ally Palmer completed a Master of Arts in the Department of Anthropology at the University of Auckland in 2012. Her research combined ethological, ethnographic and historical data to analyse inter- and intraspecies social relationships among keepers and orangutans at the zoological gardens in Auckland, New Zealand. Her doctoral research similarly employs a multidisciplinary approach to examining relationships between orangutans and humans – scientists, tourists, volunteers and workers – at a range of sites associated with transnational orangutan conservation, with the aim of assessing the efficacy and purpose of orangutan conservation from both species' perspectives.

Mandy Paterson has held the position of principal scientist with RSPCA Queensland for five years. She has a keen interest in animal welfare, policy development, exhibited animals, shelter-animal management, wildlife, pest management and many other areas within the RSPCA. She represents RSPCA Queensland on many government committees including in the area of kangaroo management, injured koala care, and Fraser Island dingoes to name a few. Mandy is also involved in research in collaboration with the University of Queensland, serves on the board of the Australia and New Zealand Council for the Care of Animals in Research and Teaching and is the chair of the Centre for Animal Welfare and Ethics Advisory Committee at the University of Queensland. Mandy worked as a clinical veterinarian for over 20 years and her PhD research explored the parent verification of dogs through blood groups and protein polymorphism.

Clive Phillips studied agriculture at Reading University and obtained a PhD in dairy cow nutrition and behaviour from the University of Glasgow. He lectured in farm animal production and medicine at the Universities of Wales and Cambridge, conducting research into cattle and sheep welfare. As the inaugural holder of the University of Queensland Chair in Animal Welfare he is now involved in research in animal welfare and ethics and the development and implementation of state and federal government animal welfare policies. He has published approximately eight books and 230 journal articles on animal welfare and management.

Amanda Stuart's art practice and research explores human relationships with the Australian natural environment. She has a particular focus on 'outsider' species. A previous science degree and work as a park ranger has been crucial in developing her fascination with non-human animal species that are reviled or perceived as dangerous by humans. Since graduating from the Australian National University School of Art with First Class Honours and the University Medal, she has exhibited her work nationally and internationally. In 2009 she participated in the inaugural Sculpture by the Sea, Denmark. Amanda has taught and researched at the sculpture workshop at the School of Art, Australian National University. Amanda's doctorate in visual arts researched the tense relations between wild dogs, dingoes and humans in southeast Australia.

Anne Taylor is an artist and academic. Since completing her PhD at the Queensland College of Art, Griffith University, she has continued to research the ethical dimensions of contemporary art through feminist perspectives, and the role of aesthetic experience in prompting awareness of environmental concerns. She has presented papers at various national and international academic conferences and been published in peer reviewed forums. Her paintings, drawings and prints depict watery realms as alternative human environments, evoking our nurturing but confined uterine beginnings and the metaphorical parallels between exotic marine organisms and human morphology. Her work also alludes to the intrusion of technology into the natural world, and to the over-exploitation of its resources.

Index

Index